Report of Investigations 9678

Results of In-Mine Research in Support of the Investigation of the Sago Mine Explosion

By Kenneth L. Cashdollar, Eric S. Weiss, Samuel P. Harteis, P.E., Michael J. Sapko, and John E. Urosek, P.E.

DEPARTMENT OF HEALTH AND HUMAN SERVICES
Centers for Disease Control and Prevention
National Institute for Occupational Safety and Health
Pittsburgh Research Laboratory
Pittsburgh, PA

September 2009

This document is in the public domain and may be freely copied or reprinted.

Disclaimer

Mention of any company or product does not constitute endorsement by the National Institute for Occupational Safety and Health (NIOSH). In addition, citations to Web sites external to NIOSH do not constitute NIOSH endorsement of the sponsoring organizations or their programs or products. Furthermore, NIOSH is not responsible for the content of these Web sites. All Web addresses referenced in this document were accessible as of the publication date.

Ordering Information

To receive documents or other information about occupational safety and health topics, contact NIOSH at

> Telephone: **1–800–CDC–INFO** (1–800–232–4636)
> TTY: 1–888–232–6348
> e-mail: cdcinfo@cdc.gov
>
> or visit the NIOSH Web site at **www.cdc.gov/niosh**.

For a monthly update on news at NIOSH, subscribe to NIOSH *eNews* by visiting **www.cdc.gov/niosh/eNews**.

DHHS (NIOSH) Publication No. 2009–168

September 2009

SAFER • HEALTHIER • PEOPLE™

CONTENTS

Page

Abstract 1
Introduction 2
Experimental facilities and instrumentation 2
Summary of explosion tests 8
    Test 1 (LLEM test #501), April 15, 2006 8
    Test 2 (LLEM test #502), June 15, 2006 17
    Test 3 (LLEM test #503), August 4, 2006 29
    Test 4 (LLEM test #504), August 16, 2006 32
    Test 5 (LLEM test #505), August 23, 2006 36
    Test 6 (LLEM test #506), October 19, 2006 55
Summary and conclusions 74
Acknowledgments 76
References 76
Appendix A.—MSHA-WVOMHS&T-NIOSH protocols for the LLEM explosion tests 78
    NIOSH-MSHA-WVOMHS&T seal testing – test No. 1 protocol – LLEM test #501 79
    NIOSH-MSHA-WVOMHS&T seal testing – test No. 2 protocol – LLEM test #502 85
    NIOSH-MSHA-WVOMHS&T seal testing – test No. 3 protocol – LLEM test #503 89
    NIOSH-MSHA-WVOMHS&T seal testing – test No. 4 protocol – LLEM test #504 94
    NIOSH-MSHA-WVOMHS&T seal testing – test No. 5 protocol – LLEM test #505 97
    NIOSH-MSHA-WVOMHS&T seal testing – test No. 6 protocol – LLEM test #506 99
Appendix B.—Seal construction descriptions 104
    1. Standard 40-in Omega block seal, 2001 design, in crosscut 2, March 23, 2006 104
    2. Hybrid 40-in Omega block seal design in crosscut 3, March 24, 2006 111
    3. 40-in Omega block seal, 2001 design, at 320 ft outby C-drift face, May 18, 2006 115
    4. Sago 40-in Omega block seal design in crosscut 3, July 5, 2006 119
    5. Sago 40-in Omega block seal design at 320 ft outby C-drift face, July 7, 2006 123
    6. Mitchell-Barrett solid-concrete-block seal in crosscut 3, September 11–15, 2006 128
    7. Sago 40-in Omega block seal design using the blocks from Sago mine at 320 ft
        outby C-drift face, September 21, 2006 132
Appendix C.—Air leakage data for seals 134

ILLUSTRATIONS

1. Plan view of the Lake Lynn experimental mine 3
2. Seal test area in the LLEM 4
3. Front and side views of pressure transducers in front of seals 5
4. LVDT mounted behind seal to measure movement 5
5. Roof plates and belt hanger at 134 ft from the face 6
6. Roof plates and belt hanger at 234 ft from the face 7
7. Setup for test 1 (LLEM #501) 9
8. Pressures and LVDT displacement at the Omega block seal in X-2 during test 1
    (LLEM #501) 10
9. Pressures and LVDT displacement at the Omega block seal in X-3 during test 1
    (LLEM #501) 11

CONTENTS—Continued

Page

10. Wood cribs at 312 ft looking inby in C-drift before test 1 (LLEM #501) 14
11. Battery charger located at ~602 ft in C-drift..14
12. Battery charger and debris from the wood cribs near X-7 after test 1 (LLEM #501) 15
13. Roof plates at various distances from the face after test 1 (LLEM #501) 16
14. Setup for test 2 (LLEM #502) 17
15. Pressures and LVDT displacement at the Omega block seal in X-2 during test 2
    (LLEM #502)..18
16. Pressures and LVDT displacement at the Omega block seal in X-3 during test 2
    (LLEM #502)..19
17. Pressures and LVDT displacement at C-drift Omega block seal during test 2
    (LLEM #502)..20
18. Remains of X-3 seal viewed from C-drift after test 2 (LLEM #502) 23
19. Postexplosion view looking outby from original seal position in C-drift...............................24
20. Postexplosion view looking outby from 403 ft after test 2 (LLEM #502) 24
21. Debris after test 2 (LLEM #502) looking outby from X-6 location 25
22. Debris after test 2 (LLEM #502) looking outby from X-7 location 25
23. Debris after test 2 (LLEM #502) looking outby from 850 ft...26
24. Final location of battery charger at ~770 ft 27
25. Roof plates at various distances from the face after test 2 (LLEM #502) 28
26. Setup for test 3 (LLEM #503)...29
27. Pressures and LVDT displacement at the X-2 seal during test 3 (LLEM #503) 30
28. Pressures and LVDT displacement at the X-3 seal during test 3 (LLEM #503) 30
29. Pressures and LVDT displacement at the C-drift seal during test 3 (LLEM #503) 31
30. Setup for test 4 (LLEM #504)...32
31. Pressures and LVDT displacement at the X-2 seal during test 4 (LLEM #504) 33
32. Pressures and LVDT displacement at the X-3 seal during test 4 (LLEM #504)...................33
33. Pressures and LVDT displacement at the C-drift seal during test 4 (LLEM #504) 34
34. Setup for test 5 (LLEM #505)...36
35. Pressures and LVDT displacement at the X-2 seal during test 5 (LLEM #505) 37
36. Pressures and LVDT displacement at the X-3 seal during test 5 (LLEM #505) 38
37. Pressures and LVDT displacement on an expanded time scale for the X-3 seal during
    test 5 (LLEM #505) 38
38. Pressures and LVDT displacement at the C-drift seal during test 5 (LLEM #505) 39
39. Pressures and LVDT displacement on an expanded time scale at the C-drift seal during
    test 5 (LLEM #505) 40
40. Seal in X-2 that survived test 5 (LLEM #505) 43
41. Debris from X-3 seal after test 5 (LLEM #505) viewed from C-drift...................................43
42. Debris from X-3 seal after test 5 (LLEM #505) looking into X-3 toward B-drift...................44
43. Debris from the X-3 seal at the X-3 stopping location after test 5 (LLEM #505).................44
44. Debris from the X-3 seal and stopping piled against the far wall of A-drift after test 5
    (LLEM #505)..45
45. Postexplosion view in C-drift looking outby toward original seal location at 320 ft...............46
46. Debris in C-drift after test 5 (LLEM #505) looking outby toward X-4 (right side of
    photo) at 355 ft..46

CONTENTS—Continued

Page

47. Debris from C-drift seal, stopping, and wood cribs looking outby from 403 ft 47
48. Debris in C-drift after test 5 (LLEM #505) looking outby from X-6 at 547 ft 47
49. Debris in C-drift after test 5 (LLEM #505) looking outby from ~600 ft 48
50. Debris in C-drift after test 5 (LLEM #505) looking outby from 757 ft 49
51. Debris in C-drift after test 5 (LLEM #505) at ~850 ft 49
52. Debris in C-drift after test 5 (LLEM #505) at ~880 ft 50
53. Final location of battery charger near 720 ft after test 5 (LLEM #505) 51
54. Roof plates at 84, 134, and 184 ft from the face after test 5 (LLEM #505) 52
55. Roof plates at 234, 304, and 403 ft from the face after test 5 (LLEM #505) 53
56. Belt hanger near DG panel at 403 ft 54
57. Belt hanger at 403 ft after test 5 (LLEM #505) 54
58. Setup for test 6 (LLEM #506) 55
59. Pressures and LVDT displacement at the X-2 seal during test 6 (LLEM #506) 56
60. Pressures and LVDT displacement at the X-3 seal during test 6 (LLEM #506) 57
61. Pressures and LVDT displacement at the C-drift seal during test 6 (LLEM #506) 58
62. Pressures and LVDT displacement on an expanded time scale at the C-drift seal during test 6 (LLEM #506) 58
63. Raw and smoothed pressure data for the horizontal transducer during test 6 (LLEM #506) 59
64. Raw and smoothed pressure data for the embedded transducer during test 6 (LLEM #506) ... 59
65. Postexplosion view in C-drift looking outby toward original seal location at 320 ft after test 6 (LLEM #506) 62
66. Postexplosion view in C-drift looking outby from X-4 at 355 ft 63
67. Debris in C-drift after test 6 (LLEM #506) looking outby from 403 ft 63
68. Debris in C-drift after test 6 (LLEM #506) looking outby from ~500 ft 64
69. Debris in C-drift after test 6 (LLEM #506) looking outby from ~600 ft 64
70. Debris in C-drift after test 6 (LLEM #506) looking outby from 757 ft 65
71. Debris in C-drift after test 6 (LLEM #506) looking outby from ~850 ft 65
72. Debris after test 6 (LLEM #506) looking outby from ~950 ft 66
73. Debris after test 6 (LLEM #506) looking outby from ~1,050 ft 67
74. Debris after test 6 (LLEM #506) looking outby from ~1,150 ft 67
75. Debris in C-drift after test 6 (LLEM #506) looking outby from ~1,240 ft 68
76. Debris in C-drift after test 6 (LLEM #506) looking outby from ~1,440 ft 68
77. Final position of battery charger after test 6 (LLEM #506) 69
78. Roof plates at 84 and 134 ft from the face after test 6 (LLEM #506) 70
79. Roof plates at 184 and 234 ft from the face after test 6 (LLEM #506) 71
80. Roof plates at 304 and 403 ft from the face after test 6 (LLEM #506) 72
81. Belt hanger near DG panel at 403 ft 73
82. Belt hanger at 403 ft after explosion test 6 (LLEM #506) 73

Standard 40-in Omega block seal, 2001 design, in crosscut 2, March 23, 2006:

B-1. BlocBond applied to concrete mine floor 106
B-2. BlocBond being applied to gaps between the Omega blocks 106
B-3. As each Omega block course was completed, BlocBond was also coated on the exposed B- and C-drift sides of the blocks 107

CONTENTS—Continued

Page

B-4. Schematic illustrating alternating courses and staggered block joints............................107
B-5. Cut blocks on top course to provide 2-in gap to roof 108
B-6. View of top portion of seal from C-drift side after B-drift side was completed...............108
B-7. One row of wood placed flush with the edge of the blocks on the B-drift side................109
B-8. View from the C-drift side showing the rough-cut lumber wedged between the mine roof and blocks across the center of the seal..109
B-9. Closeup of the wood wedges, installed between the mine roof and the rough-cut board, used to tighten the Omega block seal 110

Hybrid 40-in Omega block seal design in crosscut 3, March 24, 2006:

B-10. Dry BlocBond being spread to a 0.5-in-thick layer on the dampened concrete floor in X-3 113
B-11. Wetting the dry layer of BlocBond before positioning the Omega blocks.......................113
B-12. Positioning the first course of Omega blocks 114
B-13. Applying BlocBond to the top of the first course...114

40-in Omega block seal, 2001 design, at 320 ft outby C-drift face, May 18, 2006:

B-14. A 0.25- to 0.5-in-thick layer of properly mixed BlocBond on the floor...........................117
B-15. Omega blocks laid wet with fully mortared (BlocBond) horizontal and vertical joints......117
B-16. Completing the 11th and final full block course on the outby side 118
B-17. Hand slinging the BlocBond to fill the gaps between the previously installed center board and the outby board..118

Sago 40-in Omega block seal design in crosscut 3, July 5, 2006:

B-18. Applying the 1.5-in-thick layer of dry BlocBond to the dampened concrete floor in X-3 121
B-19. Applying a fine water spray to the dry BlocBond layer 121
B-20. Applying the properly mixed BlocBond to the first course of Omega blocks..................122
B-21. Driving the wedges (skin to skin) between the crossboard and mine roof......................122

Sago 40-in Omega block seal design at 320 ft outby C-drift face, July 7, 2006:

B-22. Applying a 1.5-in-thick layer of dry BlocBond on the dampened concrete floor across C-drift...125
B-23. Wetting the dry layer of BlocBond with a fine spray of water...125
B-24. Positioning the Omega blocks in a staggered pattern for the first course.......................126
B-25. Applying BlocBond by shovel and gloved hands to the first course 126
B-26. Positioning the second course of Omega blocks in a transverse pattern 127
B-27. Installing the crossboard on the outby side of C-drift seal 127

CONTENTS—Continued

Page

Mitchell-Barrett solid-concrete-block seal in crosscut 3, September 11–15, 2006:

B-28. Full wet-bed construction on all horizontal and vertical block joints 129
B-29. Installing cut blocks on top course 130
B-30. Completely filling the gap between the top course of blocks and the mine roof with mortar 130
B-31. Mortar filling the gaps between the steel angle hitching and the blocks along the floor and ribs 131

TABLES

1. Explosion tests of seals for MSHA during LLEM explosion research 8
2. Pressures and LVDT displacement at the seals during test 1 (LLEM #501) 12
3. Wall pressures and flame travel during test 1 (LLEM #501) 12
4. MSHA guidelines for acceptable air leakage through a seal 13
5. Pressures and LVDT displacement at the seals during test 2 (LLEM #502) 22
6. Wall pressures and flame travel during test 2 (LLEM #502) 22
7. Pressures and LVDT displacement at the seals during test 4 (LLEM #504) 35
8. Wall pressures and flame travel during test 4 (LLEM #504) 35
9. Pressures and LVDT displacement at the seals during test 5 (LLEM #505) 41
10. Wall pressures and flame travel during test 5 (LLEM #505) 42
11. Pressures and LVDT displacement at the seals during test 6 (LLEM #506) 60
12. Wall pressures and flame travel during test 6 (LLEM #506) 61
13. Summary of explosion pressures on various seals 74
C-1. Air leakage measurements before test 1 (LLEM #501) 134
C-2. Air leakage measurements after test 1 (LLEM #501) 134
C-3. Air leakage measurements before test 2 (LLEM #502) 134
C-4. Air leakage measurements after test 2 (LLEM #502) 134
C-5. Air leakage measurements before test 3 (LLEM #503) 135
C-6. Air leakage measurements after test 3 (LLEM #503) 135
C-7. Air leakage measurements after test 4 (LLEM #504) 135
C-8. Air leakage measurements after test 5 (LLEM #505) 135
C-9. Air leakage measurements before test 6 (LLEM #506) 136
C-10. Air leakage measurements after test 6 (LLEM #506) 136

ACRONYMS AND ABBREVIATIONS USED IN THIS REPORT

ASTM	American Society for testing and Materials
BDP	bidirectional probe
CFR	Code of Federal Regulations
CH_4	methane
DG	data-gathering
KS	Kinetic Systems
LVDT	linear variable displacement transducer
MSHA	Mine Safety and Health Administration
NI	National Instruments
NIOSH	National Institute for Occupational Safety and Health
LLEM	Lake Lynn Experimental Mine
PRL	Pittsburgh Research Laboratory (NIOSH)
WVOMHS&T	West Virginia Office of Miners' Health, Safety, and Training
X	crosscut (e.g., "X-1" stands for "crosscut 1")

UNIT OF MEASURE ABBREVIATIONS USED IN THIS REPORT

cfm	cubic foot per minute
ft	foot
ft^2	square foot
ft^3	cubic foot
gal	gallon
hr	hour
Hz	hertz
in	inch
in^2	square inch
in H_2O	inch of water
lb	pound
min	minute
ms	millisecond
psi	pound-force per square inch
sec	second

DEDICATION

This report was initially prepared by Kenneth L. Cashdollar and is dedicated to his memory. Ken passed away on March 4, 2009. Ken never wavered from his continuing commitment to conduct the highest-quality, solution-oriented, scientific research focused on reducing the risk of fatalities from explosions in the mining and chemical industries.

RESULTS OF IN-MINE RESEARCH IN SUPPORT OF THE INVESTIGATION OF THE SAGO MINE EXPLOSION

By Kenneth L. Cashdollar,[1] Eric S. Weiss,[2] Samuel P. Harteis, P.E.,[3] Michael J. Sapko,[4] and John E. Urosek, P.E.[5]

ABSTRACT

The Mine Safety and Health Administration (MSHA) and the West Virginia Office of Miners' Health, Safety, and Training (WVOMHS&T) investigated the explosion at the Sago Mine in West Virginia, which occurred on January 2, 2006, and resulted in 12 fatalities. As part of the investigation, the agencies requested that the National Institute for Occupational Safety and Health's (NIOSH) Pittsburgh Research Laboratory evaluate the effects of explosions on specific mine ventilation seals and other structures and objects at its Lake Lynn Experimental Mine (LLEM). The results of the LLEM study would assist MSHA and WVOMHS&T in a more thorough understanding of the various questions that arose during their investigations of the explosion. Six large-scale explosion tests were conducted in the LLEM from April to October 2006. The protocols for these tests, and in particular the procedures for constructing various Omega block seals, were developed mainly by MSHA and WVOMHS&T. NIOSH developed the experimental procedures at the LLEM that would provide the required range of explosion pressures against the seals. Three 40-in-thick seal designs using Omega 384 low-density block were constructed in the LLEM and exposed to various explosion pressures. These seal designs are referred to in this report as the "2001 design," the "hybrid design," and the "Sago design."

The 2001-design Omega block seal (80 in high) located in crosscut 2 of the LLEM survived all six explosions, with pressure loadings up to 51 psi. The 2001-design Omega block seal (88 in high) in C-drift was destroyed during Test 2, which subjected the seal to a head-on explosion that resulted in a pressure loading of 51 psi. The height difference between the two seals and the orientation of each seal to the explosion were contributing factors. The higher seal would be weaker for the same seal thickness. The hybrid Omega block seal in crosscut 3 survived an explosion at a pressure loading of 25 psi and failed during another explosion at a pressure loading of 39 psi at the seal. Based on these tests, it seems that the hybrid seal design is weaker than the 2001 seal design. The Sago Omega block seals were constructed in crosscut 3 and C-drift before Test 3. The crosscut 3 seal survived a pressure loading of 18 psi and was destroyed during an explosion at a pressure loading of 35 psi at the seal. The C-drift seal survived a head-on explosion that resulted in a pressure loading of 21 psi and was destroyed during an explosion with a pressure loading of 57 psi

---

[1] Principal Research Physical Scientist, Pittsburgh Research Laboratory, National Institute for Occupational Safety and Health, Pittsburgh, PA (deceased).
[2] Team Leader (Senior Research Mining Engineer), Lake Lynn Laboratory Section, Pittsburgh Research Laboratory, National Institute for Occupational Safety and Health, Pittsburgh, PA.
[3] Research Mining Engineer, Pittsburgh Research Laboratory, National Institute for Occupational Safety and Health, Pittsburgh, PA.
[4] Principal Research Physical Scientist, Pittsburgh Research Laboratory, National Institute for Occupational Safety and Health, Pittsburgh, PA (retired).
[5] Chief, Mine Emergency Operations, Pittsburgh Safety and Health Technology Center, Mine Safety and Health Administration, Pittsburgh, PA.

at the seal. Based on these LLEM tests, it seems that the Sago seal design is weaker than the 2001 seal design, yet it still complied with the requirements of 30 CFR 75.335(a)(2) that were in effect at the time of the mine disaster.

During these LLEM explosion tests, the distance of seal debris travel was also measured. In Test 5, the C-drift seal was destroyed during a head-on explosion that resulted in a pressure loading of 57 psi, and the seal debris was thrown over 500 ft. In Test 6, the C-drift seal was destroyed during a head-on explosion that resulted in a pressure loading of 93 psi, and the Omega block debris was thrown over 900 ft. During these tests, the explosion pressure effects on other structures and objects were also documented. Unless otherwise specified, all of the pressure data listed in this report represent the highest recorded smoothed (averaged over 10 ms) explosion pressure loadings.

The information in this report was used by MSHA and WVOMHS&T as supporting data in their analyses and investigative reports of the Sago Mine explosion.

INTRODUCTION

At the request of the Mine Safety and Health Administration (MSHA) and the West Virginia Office of Miners' Health, Safety, and Training (WVOMHS&T), the National Institute for Occupational Safety and Health's (NIOSH) Pittsburgh Research Laboratory (PRL) evaluated the effects of explosions on specific mine ventilation seals at its Lake Lynn Experimental Mine (LLEM) to assist the agencies in their investigations of the explosion at the Sago Mine in West Virginia, which occurred on January 2, 2006, and resulted in 12 fatalities. Six full-scale explosion tests were conducted in the LLEM from April to October 2006 to help answer questions regarding possible scenarios for the Sago explosion. The protocols for these tests, and in particular the procedures for constructing various Omega block seals, were developed mainly by MSHA and WVOMHS&T. NIOSH developed the experimental procedures at the LLEM that would provide the required range of explosion pressures against the seals. NIOSH also documented the seal construction, determined and installed the appropriate instrumentation, conducted the explosion tests, analyzed the data, and photographically documented the postexplosion observations. By comparing the results of known explosion loading pressures on the various ventilation structures and objects in the LLEM with their observations at the Sago Mine, MSHA and WVOMHS&T could better analyze the explosion pressures that may have occurred at the Sago Mine.

EXPERIMENTAL FACILITIES AND INSTRUMENTATION

The full-scale explosion tests were conducted in the LLEM [Mattes et al. 1983; Triebsch and Sapko 1990] (Figure 1). The LLEM is a former limestone mine, and five new drifts (horizontal passageways in a mine) were developed in 1979–1980 to simulate the geometries of modern U.S. coal mines. The mine has four parallel drifts: A, B, C, and D. D-drift is a 1,640-ft-long single entry that can be separated from E-drift by an explosion-proof bulkhead door. To simulate room-and-pillar workings, drifts A, B, and C can be used. These three drifts are each approximately 1,600 ft long, with seven crosscuts at the inby end. An explosion-resistant bulkhead door is used to separate the multiple entries from E-drift (Figure 1). Drifts C and D are connected by E-drift, a 500-ft-long entry that simulates a longwall face. An 8-in-thick reinforced concrete floor runs the entire length of each of these drifts and crosscuts. The explosion tests were conducted in the multiple-entry area of A-, B-, and C-drifts. The entries are about 20 ft wide by about 6.5 ft high, with cross-sectional areas of 130–140 ft^2. The LLEM bulkhead door and some of the other infrastructure were designed to withstand explosion overpressures of up to 100 psi. Higher pressures have been recorded at areas away from these structures.

Each LLEM drift has 10 data-gathering (DG) stations inset in the rib wall at the locations shown in Figures 1 and 2. Each DG station houses a strain gauge transducer to measure the pressure and an optical sensor to detect the flame arrival. The explosion pressure is dynamic in nature and is composed of two components: a quasi-static pressure component (the pressure that is exerted in all directions and measured perpendicular to the gas flow) and a wind or velocity pressure component (pressure due to gas flow). The total explosion pressure is the sum of the quasi-static pressure and the wind or velocity pressure. The transducers in the DG stations in the wall, which is perpendicular to the gas flow, measure the quasi-static pressures. All of the explosion pressures presented in this report are an averaged overpressure or gauge pressure (pressures above local atmospheric pressure) rather than absolute pressures. In the past, Nagy [1981, p. 58] referred to the quasi-static pressure as the "static pressure" to differentiate it from the "dynamic pressure," or velocity component. However, the quasi-static pressure is not actually static as it does vary with time during the explosion. Zipf et al. [2007] provide a more detailed discussion on explosion pressures. Other instruments may also be installed at various locations in the LLEM during an explosion test.

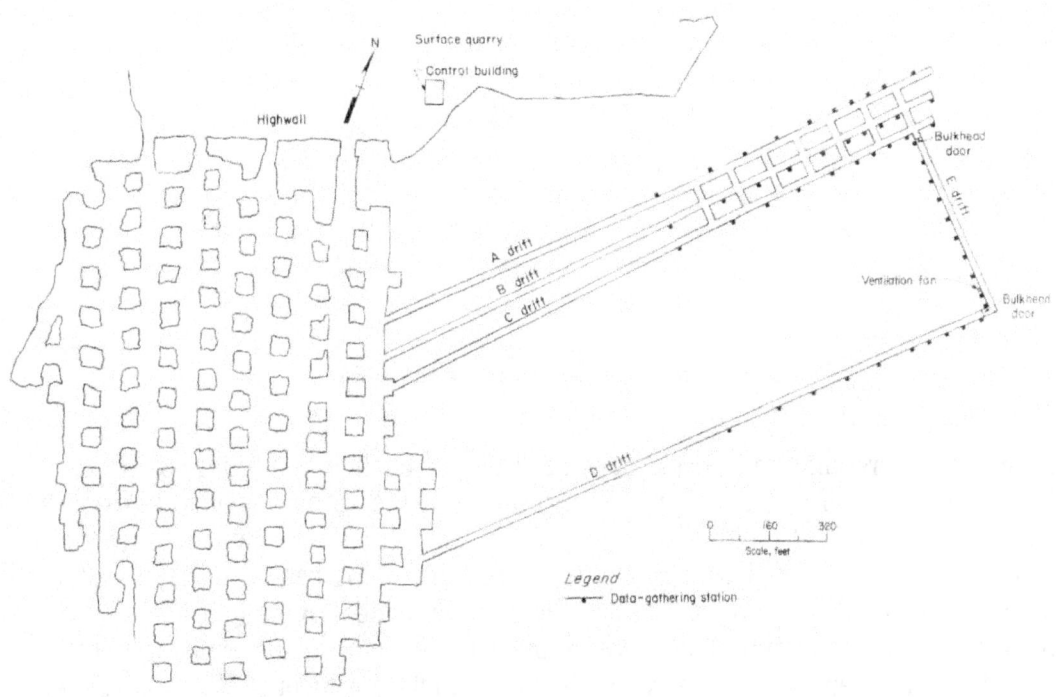

Figure 1.—Plan view of the Lake Lynn Experimental Mine.

During the normal course of underground coal mining, it sometimes becomes necessary to install seals to isolate abandoned or worked-out areas of a mine. Since 1992, 30 CFR[6] 75.335 required a seal to "withstand a static horizontal pressure of 20 pounds per square inch." This regulation formed the basis for previous PRL evaluations of explosion-resistant seals at the LLEM [Stephan 1990a,b; Greninger et al. 1991; Weiss et al. 1993a,b,c; 1996; 1997; 1999]. During the 1990s, PRL and MSHA jointly evaluated the capability of various seal construction materials and designs to meet or exceed the requirements of the CFR.

---

[6] *Code of Federal Regulations.* See CFR in references.

Figure 2 is a closeup plan view of the seal test area in the multiple-entry area of the LLEM. In this example, there are seals in the first four crosscuts from the face or closed end of C-drift. Note that, in the LLEM, the first crosscut ("1" in Figure 2) is the one nearest the face. The crosscuts are 17–19 ft wide by ~7.2 ft high with a cross-sectional area of about 130 ft^2. Explosion-resistant seals from a previous study were located in X-1[7] and X-2. The flammable CH_4-air gas zone is at the face (closed end) of C-drift. The gas zone is confined on the outby end by a plastic diaphragm. The bulkhead door is closed between C- and E-drifts before the test. For an explosion test, the gas is ignited and the explosion pressure travels out C-drift.

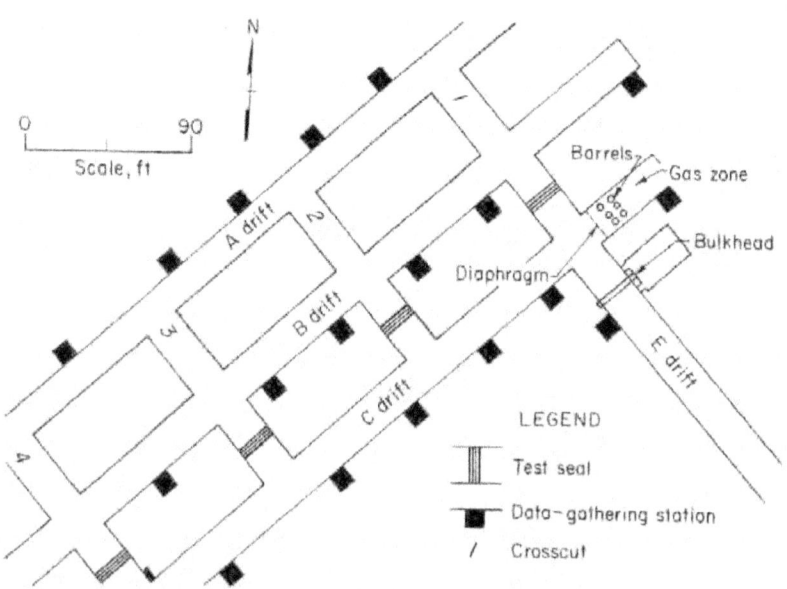

Figure 2.—Seal test area in the LLEM.

Examples of pressure transducers in front of a seal are shown in Figure 3. For the first two tests (LLEM #501–502) of the series, there was at least one pressure transducer in front of each seal. For the later tests (LLEM #503–506), there were generally at least two pressure transducers in front of each seal—one mounted horizontally to face the incoming head-on pressure wave (to measure the total explosion pressure at that location) and one mounted vertically to be perpendicular to the incoming pressure wave (to measure the quasi-static explosion pressure at that location). Behind each seal was a linear variable differential transducer (LVDT), as shown in Figure 4. The LVDT measures the movement of the seal [Weiss et al. 1999, pp. 5–6]. Also shown in Figure 4 are the horizontal and vertical yellow breakwires used to measure the time of seal failure. During the explosion tests, a high-speed, PC-based National Instruments (NI) data acquisition system collected the data from the various instruments at a sampling rate of 1,500 samples per second. The reported data were normally averaged over 10 ms (15-point smoothing). For some of the tests, a second Kinetic Systems (KS) data acquisition system collected the data at 5,000 samples per second.

---

[7]The abbreviation "X" stands for "crosscut" throughout this report, e.g., "X-1" stands for "crosscut 1."

 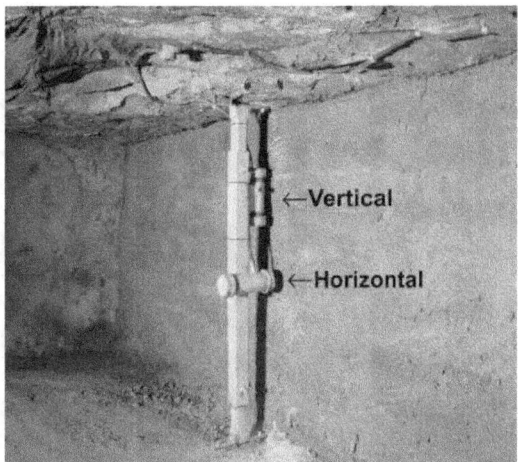

Figure 3.—Front and side views of pressure transducers in front of seals.

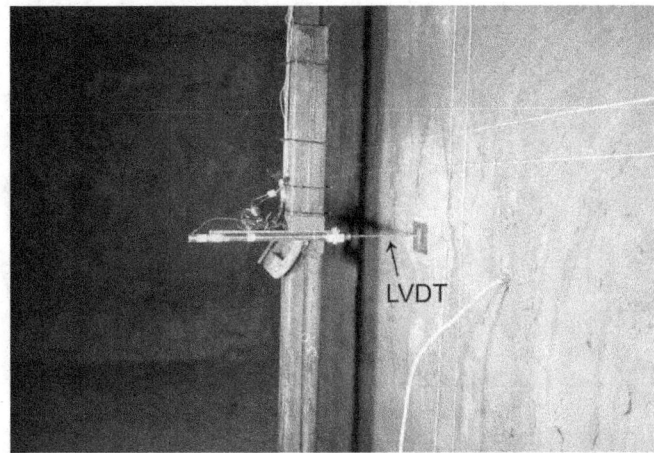

Figure 4.—LVDT mounted behind seal to measure movement.

Typical examples of the round and square roof plates and belt hangers that were attached to the LLEM roof at various distances from the face are shown in Figures 5-6. The large round plates are ~19 inches in diameter and the metal is 0.033 in thick. The large square plates (also known as spider plates) are ~17 in by 17 in and the metal is 0.038 in thick. A small square bearing plate is below each large round or square plate in the photos. The small square bearing plates are ~8 in by 8 in by 0.146 in thick. Unless specified otherwise, all subsequent discussion in this report will refer to the large square plates and not the small square plates. The belt hangers at the bottom of Figures 5-6 are ~4 in by 4 in on each side by 0.25 in thick.

For this series of LLEM explosion tests, MSHA hired a certified surveyor (Allegheny Surveys, Inc., Birch River, WV) to survey the positions of the various objects in the mine before the explosions and to survey the debris from seals and stoppings, wood cribs, and various other objects after the explosions.

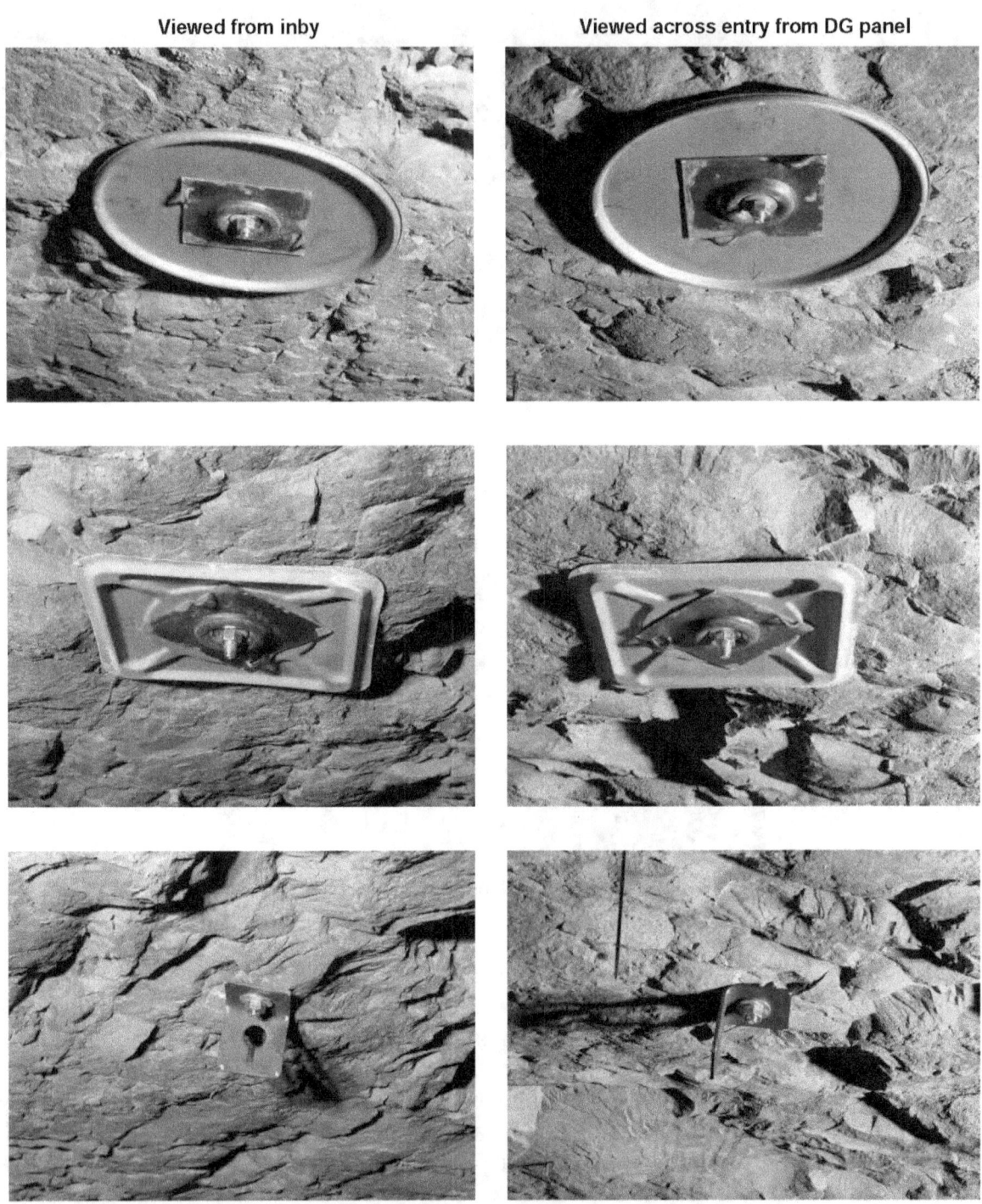

Figure 5.—Roof plates and belt hanger at 134 ft from the face.

Figure 6.—Roof plates and belt hanger at 234 ft from the face.

SUMMARY OF EXPLOSION TESTS

A summary of the LLEM explosion tests conducted in 2006 is presented in Table 1. Each test was given a number (1 through 6) corresponding to the test series and a number (LLEM tests #501–506) corresponding to the sequential LLEM test numbering system in place since 1982. The first three columns of the table list the current test series number, the date, and the LLEM test number. The next two columns list the location and type of mine ventilation seal. The construction details for the seals are contained in Appendix B and will be referred to when the seals are discussed in subsequent sections of this report. The next column lists the peak total explosion pressure loading (averaged or smoothed over 10 ms) from the horizontal transducer located at the middle front of each seal. The last column lists whether the seal survived or failed during the explosion. The averaged peak total explosion pressure loading is the highest smoothed value from either the NI or KS data acquisition system. The pressure values are rounded to the nearest integer.

Table 1.—Explosion tests of seals for MSHA during LLEM explosion research

Test	Date	LLEM test #	Seal Location	Seal Type	Averaged peak pressure, psi	Result
1	Apr. 15, 2006	501	X-2	2001-design 40-in Omega	23	Survived
			X-3	Hybrid 40-in Omega	25	Survived
2	Jun. 15, 2006	502	X-2	2001-design 40-in Omega	22	Survived
			X-3	Hybrid 40-in Omega	39	Failed
			C-320ft[1]	2001-design 40-in Omega	51	Failed
3	Aug. 4, 2006	503	X-2	2001-design 40-in Omega	13	Survived
			X-3	Sago-like 40-in Omega	16	Survived
			C-320ft[1]	Sago-like 40-in Omega	17	Survived
4	Aug. 16, 2006	504	X-2	2001-design 40-in Omega	15	Survived
			X-3	Sago-like 40-in Omega	18	Survived
			C-320ft[1]	Sago-like 40-in Omega	21	Survived
5	Aug. 23, 2006	505	X-2	2001-design 40-in Omega	26	Survived
			X-3	Sago-like 40-in Omega	35	Failed
			C-320ft[1]	Sago-like 40-in Omega	57	Failed
6	Oct. 19, 2006	506	X-2	2001-design 40-in Omega	51	Survived
			X-3	Solid concrete block	49	Survived
			C-320ft[1]	Sago-like 40-in Omega[2]	93	Failed

[1] C-drift, 320 ft from the closed end.
[2] This seal was constructed with actual unused Omega blocks from the Sago Mine.
NOTE: A solid-concrete-block seal was installed in X-1 for all tests.

Test 1 (LLEM Test #501), April 15, 2006

For the first test, a solid-concrete-block seal was constructed in X-1, the crosscut closest to the face; an Omega block seal based on the 2001 design [Sapko et al. 2004] was constructed in X-2; and a hybrid Omega block seal was constructed in X-3 (Figure 7). The solid-concrete-block seal in X-1 had a 16-in-thick main wall with an interlocked 32-in pilaster in the center. The 6-in by 8-in by 16-in solid concrete blocks were laid in a transverse pattern (staggered joints), and high-strength mortar was applied to all of the block-to-block interfaces as well as the block-to-strata interfaces. The mortar used was BlocBond (product No. 1225-51), a fiber-reinforced surface-bonding cement made by Quikrete Co., Atlanta, GA (50-lb average weight per bag). The gap between the top block course and the mine roof was completely packed with the BlocBond mortar; no wedges were used during construction of this seal. Keying of the seal was simulated by bolting steel angle to the ribs and floor on both sides of the seal.

The X-2 nominal 40-in-thick Omega block seal was based on the 2001 design. The X-2 seal was built using Omega 384 blocks (manufactured by Burrell Mining Products in Bluefield, WV), with nominal block dimensions of 8 in by 16 in by 24 in. No pilaster or keying was required with this seal design. The block course was alternated to stagger the block joints from front to back and left to right. The properly mixed BlocBond mortar was applied at all of the block-to-block interfaces and the block-to-strata interfaces, including the floor. Three rows of 1-in-thick by 8-in-wide hardwood rough-cut boards were run lengthwise (rib to rib) on the top course of blocks. One row of board was placed across the middle of the seal, and one row of board was placed flush with each edge of the seal. Wood wedges were driven perpendicular to these boards to tighten between the mine roof and the boards, then all gaps between the top course and mine roof were completely packed with BlocBond. Both faces of the seal were then coated with BlocBond.

The nominal 40-in-thick hybrid Omega block seal in X-3 was constructed in a manner similar to the X-2 seal except that dry BlocBond was applied to a 0.25-in depth on the floor across the width of the crosscut before any blocks were laid. This dry BlocBond layer was then dampened with water sprays. The entire first course of Omega blocks was spray-dampened with water on the bottom of each block and then positioned on the dampened bed of dry BlocBond before any additional BlocBond mortar was applied. The properly mixed BlocBond mortar was then applied by gloved hand across the top of this first course and forced as much as possible with gloved hands into the vertical block joints. The remaining courses were installed in a similar manner, except the bottoms of the blocks were not dampened, the BlocBond was applied by gloved hand to the block joints before placement of each block, and the blocks in each subsequent course were alternated to stagger the block joints. All of the block work was completed before installing the 1-in-thick by 6-in-wide hardwood boards. Wood wedges were used between the boards and the mine roof. Additional details of the seal construction procedures can be found in Appendix A under "Test No. 1 Protocol" and in Appendix B, sections 1 and 2. Details of the test procedures can also be found in Appendix A under "Test No. 1 Protocol."

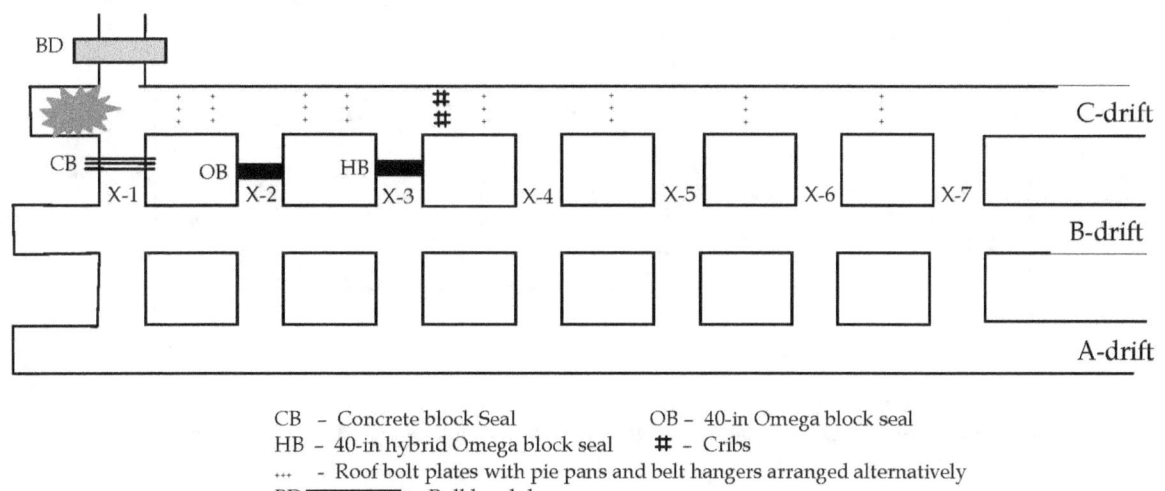

Figure 7.—Setup for Test 1 (LLEM #501).

For Test 1 (LLEM #501), a gas ignition zone was confined by a plastic diaphragm across C-drift at 47 ft from the face (Figures 2 and 7). About 661 ft^3 of natural gas was injected into the ignition zone to produce a mixture of ~10% CH_4 in air. The flammable gas was ignited at the face and the pressure pulse propagated out C-drift past the seals in X-1, X-2, and X-3, as shown in Figure 7. Note that the depth of the seals in the crosscuts is not to scale in Figure 7 or subsequent test schematics in this report. The seals in the crosscuts experienced the sweeping explosion pressure loading, i.e., the seals in the crosscuts were subjected to a nonuniform pressure loading as the explosion pressure wave traveled down C-drift past each crosscut seal location. This sweeping explosion pressure loading test method is the same method used during the LLEM seal tests during the 1990s [Stephan 1990a,b; Greninger et al. 1991; Weiss et al. 1993a,b,c; 1996; 1997; 1999]. The pressure loading on a crosscut seal will vary across the face of the seal, but the measurements in this report were taken only from the geometric center of the seal, i.e., the pressures can be significantly higher or lower near the crosscut ribs depending on the direction of the explosion travel and the various pressure wave interactions at these areas.

In this test, all of the seals survived the explosion. Figure 8 shows the pressure loading at the seal in X-2 along with the quasi-static pressures at the wall of C-drift inby (134 ft) and outby (184 ft) the seal. The pressure transducer at the seal was located approximately in the middle front of the seal at 156 ft from the face of C-drift. All of the data were averaged over 10 ms. The pressure loading at the seal was 23 psi. The curve at the top of the figure shows the seal displacement as measured by the LVDT on the back or B-drift side of the seal. The maximum recorded movement during the explosion was 0.03 in. Since the seal did not fail, there was no change in the breakwire signal.

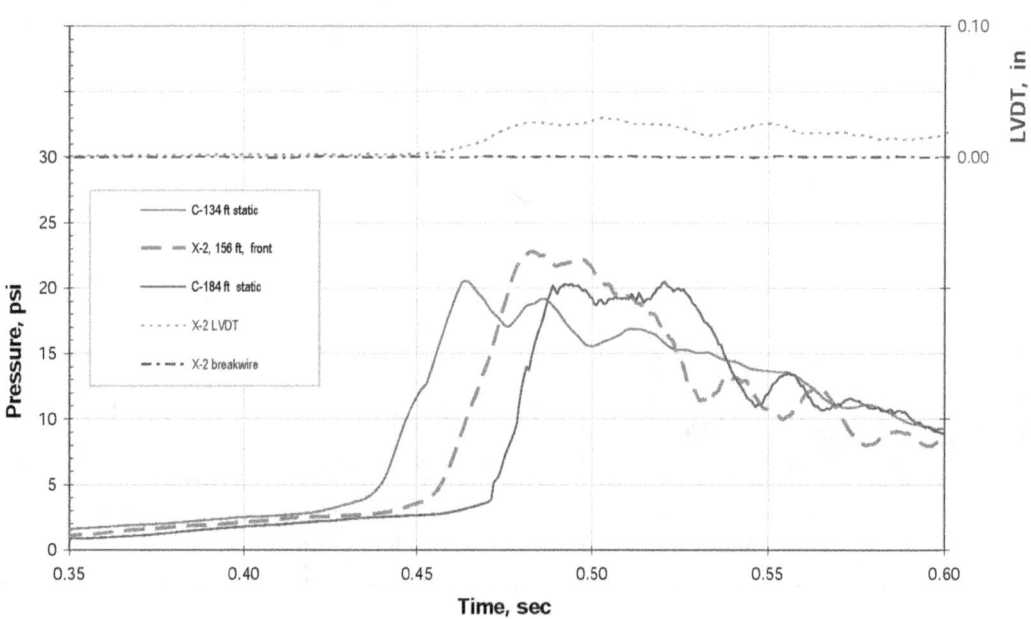

Figure 8.—Pressures and LVDT displacement at the Omega block seal in X-2 during Test 1 (LLEM #501).

Figure 9 shows the pressure loading at the seal in X-3 along with the quasi-static pressure at the wall of C-drift inby (234 ft) and outby (304 ft) the seal. The pressure transducer at the seal was located approximately in the middle front of the seal at 256 ft from the face of C-drift. All of the data were averaged over 10 ms. The pressure loading at the seal was 25 psi. The curve at the top of the figure shows the seal displacement as measured by the LVDT on the back of the seal. The maximum recorded movement during the explosion was 0.12 in. Since the seal did not fail, there was no change in the breakwire signal.

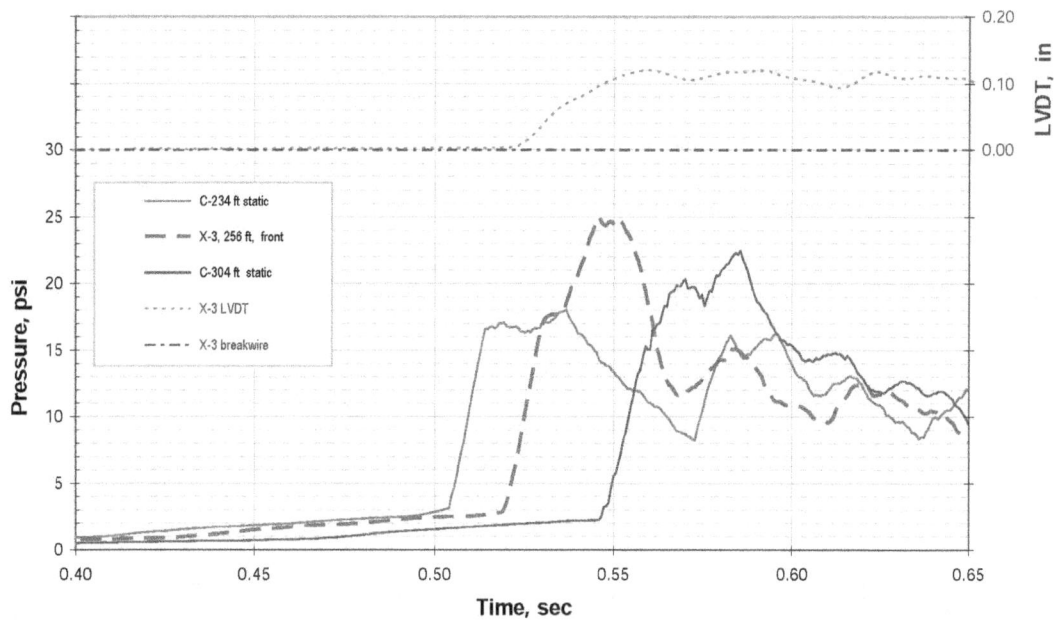

Figure 9.—Pressures and LVDT displacement at the Omega block seal in X-3 during Test 1 (LLEM #501).

Pressures, LVDT displacement, and breakwire summary data for the seals in X-2 and X-3 during Test 1 (LLEM #501) are shown in Table 2. The first column of the table lists the seal location and distance from the face of C-drift. The next two columns list the peak pressure loading at the front middle of the seals as measured by the NI and KS data acquisition systems. The data were averaged over 10 ms for all of the listed values. The next two columns list the deflection or displacement of the middle of the seal, measured in inches. The next column would list the time when the wires attached to the seals would break. However, during this test, both seals survived and the wires did not break.

Table 3 lists quasi-static pressure and flame sensor data at the various DG panels on the walls of B- and C-drifts during LLEM #501. The data were averaged over 10 ms for all of the listed values. The first three columns of the table list the position and quasi-static pressures at the DG panels in B-drift. Note that the listed distances were measured from the face of each drift. The next column lists the DG panel numbers and depicts the seal and crosscut locations. The two Omega block seals in X-2 and X-3 are depicted by blue shading. The next three columns list the position and quasi-static pressures at the DG panels in C-drift. The pressures remained relatively constant in the region of the seals in the first three crosscuts. The pressure increased somewhat at ~304 ft because of the restriction due to two wood cribs at that location. The pressures then decreased as the pressure pulse passed the open crosscuts. The last two columns in the table list the flame signal and arrival time at each of the DG panels. For this test, the flame went past the 184-ft panel but did not reach the 234-ft panel. Therefore, the interpolated flame travel distance was about 210–220 ft. The uncertainty of the flame travel distance is about ±20 ft. Note that the interpolation takes into account both the strength and duration of the flame signal at the 184-ft DG panel. Based on the initial gas zone length of 47 ft, the expansion ratio would be about 4.6 (215 ft/47 ft).

Table 2.—Pressures and LVDT displacement at the seals during Test 1 (LLEM #501)

Seal	Seal pressures psi (NI)	Seal pressures psi (KS)	LVDT deflection, in	Break time, sec	Seal	Seal pressures psi (NI)	Seal pressures psi (KS)	LVDT deflection, in	Break time, sec
X-2 BC 156 ft	22.8	22.1	0.03	—					
		Seal survived							
X-3 BC 256 ft	25.0	25.2	0.12	—					
		Seal survived							
					C-drift	*No seal in C-drift for this test*			

Table 3.—Wall pressures and flame travel during Test 1 (LLEM #501)

B-drift quasi-static pressures				C-drift quasi-static pressures			Flame signal	
Distance, ft	psi (NI)	psi (KS)		Distance, ft	psi (NI)	psi (KS)	Volts	sec (NI)
10	5.0	5.2	1 / X-1	13	—		>5	0.218
108	4.0	3.9	2	84	20.1	20.6	>5	0.396
158	~3	~3	3 / X-2	134	20.5	20.4	>5	0.432
211	3.4	3.4	4	184	20.5	21.8	>5	0.523
257	3.1	3.0	5 / X-3	234	18.0	17.5	~0	
329	5.1	5.0	6 / X-4	304	22.5	22.0	~0	
427	5.2	5.1	7 / X-5	403	14.4	12.9	~0	
526	6.5	6.7	8 / X-6	501	9.1	8.2	~0	
626	5.9	6.3	9 / X-7	598	6.1	5.8	~0	
782	4.4	4.5	10	757	3.9	3.9	~0	
			11	1,506	3.1	3.2		

An important factor to consider for any seal design is its ability to minimize air leakage through the seal. Measurements of the air leakages across the various seals were conducted before and after each of the explosion tests. For these air leakage tests, the D-drift bulkhead door (Figure 1) was closed. This directed all of the ventilation flow (from a vertical air shaft in E-drift) toward C-drift. A wooden framework with brattice cloth or curtain was erected across C-drift outby the last seal position. This curtain effectively blocked the ventilation flow, which resulted in a pressurized area on the C-drift side of the seals. By increasing the speed of the four-level LLEM main ventilation fan while in the blowing mode, we increased the pressure exerted on the structures from approximately 1 in H_2O pressure for the lowest fan speed setting to nearly 5.0 in H_2O for the highest setting. On the B-drift side of each seal, a diaphragm of brattice cloth was installed across each crosscut. A vane anemometer was used to monitor the airflow through an opening on each brattice to determine the leakage rates through each seal while the differential pressure across the seal was measured.

Based on data previously collected during the early LLEM seal testing program [Stephan 1990a; Greninger et al. 1991], MSHA developed guidelines for acceptable air leakage rates through seals for use during the LLEM seal evaluation programs. Table 4 lists these maximum acceptable air leakage rates as a function of pressure differential. For pressure differentials up to 1 in H_2O, air leakage through the seal must not exceed 100 cfm. For pressure differentials over 3 in H_2O, air leakage must not exceed 250 cfm. The flow rate was calculated from the linear air speed measured by the vane anemometer and the area of the opening through the brattice cloth behind each seal.

The air leakages of the two seals in X-2 and X-3 were measured after LLEM #501. Both seals passed the air leakage test, as shown in Table C-2 in Appendix C of this report.

Table 4.—MSHA guidelines for acceptable air leakage through a seal

Pressure differential, in H_2O	Air leakage rate, cfm
<1.0	<100
1.0 < 2.0	<150
2.0 < 3.0	<200
>3.0	<250

In addition to the seals evaluated during the first explosion test, two wood cribs were installed at 312 ft from the face in C-drift before the first test (Figure 10). These were standard cribs made from 6-in by 5-in by 30-in hardwood timbers. The individual blocks were labeled by letter and number, with crib A on the right and crib B on the left, looking inby. Between the two cribs is a bidirectional probe (BDP) at ~306 ft (as measured from the closed end of C-drift) that measured both total and dynamic explosion pressures at that location. In the background is the yellow brattice used during the preexplosion air leakage test of the seals. This brattice was removed for the explosion test and later reinstalled for the postexplosion leakage test.

During the explosion, the pressure at the crib location was 29 psi and the dynamic pressure was 9 psi. Both cribs were destroyed during the explosion, and wood debris traveled up to 880 ft from the original crib location.

Figure 10.—Wood cribs at 312 ft looking inby in C-drift before Test 1 (LLEM #501).

A battery charger from the Sago Mine had been obtained by WVOMHS&T and was placed at ~602 ft from the face in C-drift, as shown on the right looking outby in Figure 11. To the left of the battery charger is another BDP mounted vertically from the roof at 604 ft to record the total and dynamic pressures at that location during the explosion. The battery charger weighed 1,560 lb and was 40 in wide by 22.5 in high by 90.5 in long.

Figure 11.—Battery charger located at ~602 ft in C-drift.

The recorded explosion pressure at the battery charger location was 6–7 psi, and the dynamic pressure was 0.8 psi. The length of the charger was 7.5 ft. The explosion pressure would equalize from the inby to the outby end of the charger at the speed of sound. Therefore, the pressure would equalize in about 7 ms. If the pressure pulse rose from about 0 psi to its maximum value in less than 7 ms, then the inby end of the charger would have been subjected to a differential pressure that was equal to the maximum explosion pressure at that location. For Test 1, the explosion pressure loading at this location rose to its maximum value in less than 7 ms. Therefore, the inby end of the charger was subjected to a pressure loading of 6–7 psi while the outby end was still at ~0 psi. The dimensions of the end of the charger were 40 in wide by 22.5 in high, giving a cross-sectional area of 900 in^2. The ~6.5-psi pressure loading on that area would result in a total force of ~5,900 lb on the charger for a few milliseconds. The 0.8-psi dynamic pressure would continue to act on the charger for a longer time period. The battery charger moved 22 ft during Test 1. The final charger position and some of the wood debris from the cribs are shown in Figure 12.

Figure 12.—Battery charger and debris from the wood cribs near X-7 after Test 1 (LLEM #501).

Some of the roof plates were damaged during the explosion, as shown in Figure 13. On the left side of the figure, the plates are shown from inby of the plates, looking outby. On the right side of the figure, the plates are viewed looking across the entry from the DG-panel side. Four of the round plates were severely bent during the explosion. The only square plate that was bent during the explosion was the one at 304 ft. There was no obvious damage to the other round plates, square plates, or belt hangers.

Viewed from inby Viewed across entry from DG panel

134 ft from face

184 ft from face

234 ft from face

304 ft from face

Figure 13.—Roof plates at various distances from the face after Test 1 (LLEM #501).

Test 2 (LLEM Test #502), June 15, 2006

The solid-concrete-block seal in X-1, the 2001-design Omega block seal in X-2, and the hybrid Omega block seal in X-3 had survived the previous LLEM explosion test and were left in place for the second test. A new 2001-design Omega block seal was installed across C-drift at about 320 ft from the face. The details of the test procedure and seal description can be found in Appendix A under "Test No. 2 Protocol." Additional details of the C-drift seal construction can be found in Appendix B, section 3. A schematic of the test setup is shown in Figure 14.

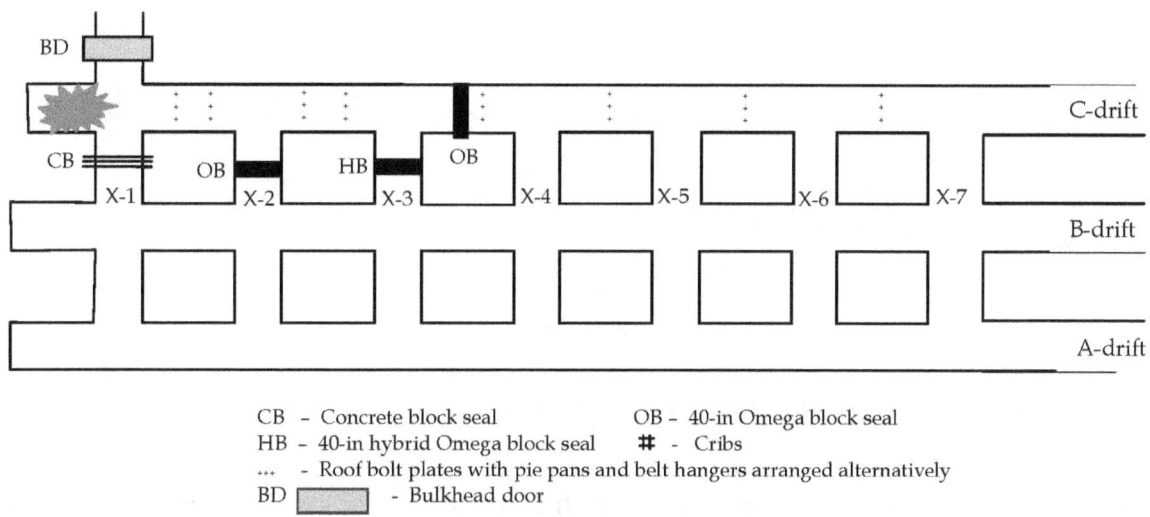

CB – Concrete block seal OB – 40-in Omega block seal
HB – 40-in hybrid Omega block seal # – Cribs
... – Roof bolt plates with pie pans and belt hangers arranged alternatively
BD – Bulkhead door

Figure 14.—Setup for Test 2 (LLEM #502).

For Test 2 (LLEM #502), a 47-ft-long, ~10% CH_4-air zone was ignited at the face of C-drift. The explosion pressure pulse propagated out C-drift past the seals in X-1, X-2, and X-3. However, in this test the explosion was confined by the seal at 320 ft in C-drift, and the pressures built up higher than in the first test. The seals in the crosscuts experienced the sweeping total explosion pressure loading, whereas the seal in C-drift experienced the head-on explosion, i.e., when the explosion pressure wave impacted the seal constructed across C-drift head on, it resulted in a uniform pressure loading across the entire seal and generated a near instantaneous and much higher reflected pressure loading on that seal.

In this test, the Omega block seal in X-2 survived, but the seals in X-3 and C-drift were destroyed by the higher pressures. Figure 15 shows the pressure loading at the seal in X-2 along with the quasi-static pressure at the wall of C-drift inby (134 ft) and outby (184 ft) the seal. The pressure transducer at the seal was located approximately in the middle front of the seal at 156 ft from the face of C-drift. All of the data were averaged over 10 ms. The pressure loading at the seal was 22 psi, similar to the value during the first test. The curve at the top of the figure shows the seal displacement as measured by the LVDT on the back of the seal. The maximum recorded movement during the explosion was 0.03 in. Since the seal did not fail, there was no change in the breakwire signal.

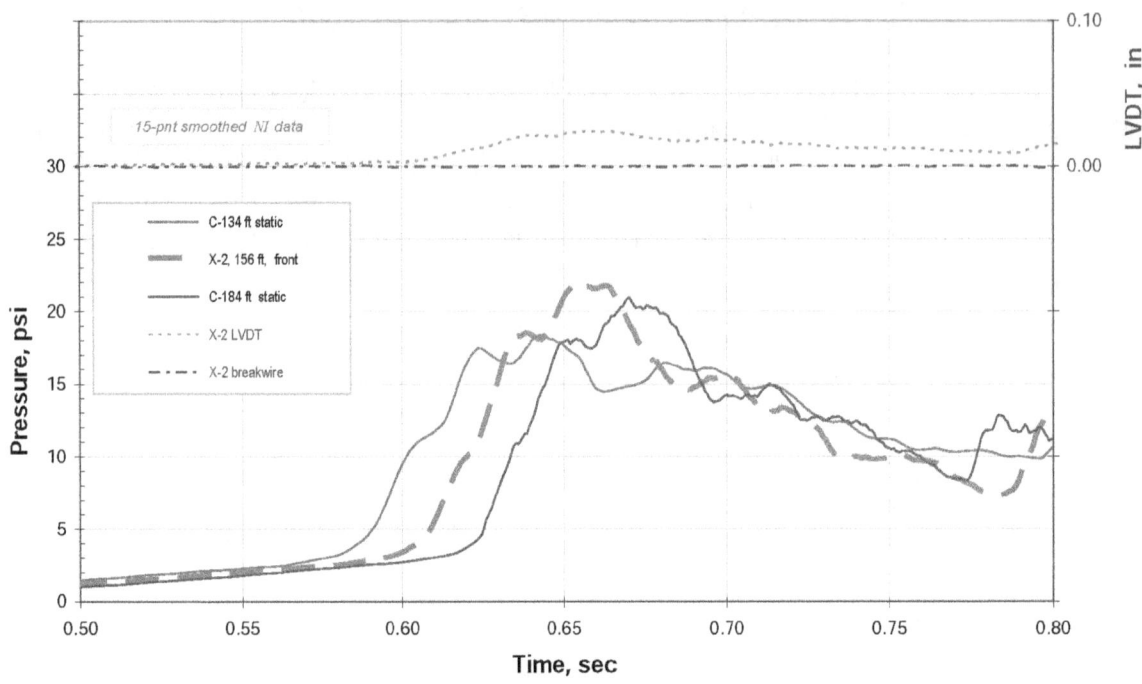

Figure 15.—Pressures and LVDT displacement at the Omega block seal in X-2 during Test 2 (LLEM #502).

The pressure loading at the seal in X-3 (with 10-ms smoothing) is shown in the top graph of Figure 16. The bottom graph of Figure 16 shows both the raw data at 1,500 samples per second and the smoothed data (10-ms average). The average pressure loading at the seal was 38.5 psi. (Note that the pressure values in the text are rounded to the nearest 0.5 psi.) The peak pressure loading from the raw data was 51 psi, but it lasted less than 1 ms. (See the section entitled "Test 6, LLEM Test #506, October 19, 2006" for further discussion of the relevance of the higher-frequency oscillations in the pressure readings.) The curve at the top of the figure shows the seal displacement as measured by the LVDT on the back of the seal. The maximum recorded movement during the explosion was >6 in as the seal was destroyed. This 6-in displacement was the maximum that the LVDT could measure. The breakwire data at the top of the figure are shown as the dot-dashed curve that drops below the baseline; the units are arbitrary. The breakwire signal showed that the wire broke, but it did not respond as quickly as it was supposed to. This was probably due to the lack of a resistor in the circuit. The LVDT and breakwire data show that the seal survived the initial outgoing pressure loading of ~25 psi at ~0.71 sec. The seal was destroyed by the subsequent higher pressure loading (~38.5 psi) that was reflected back from the C-drift seal at ~0.80 sec.

The top graph of Figure 17 shows the pressure loading at the middle front of the seal at 320 ft in C-drift and the quasi-static pressure loading as measured perpendicular to the entry near the rib at the edge of the seal. Also shown are the quasi-static pressure at the wall of C-drift at 304 ft and the total explosion pressure measured by the BDP at 306 ft. All of the data were averaged over 10 ms. The bottom graph of Figure 17 only shows the data from the two pressure transducers at the seal. The smoothed pressure loading at the seal was 49.5 psi at the front center and 51 psi at the rib. The peak pressure loadings from the 1,500-Hz raw data (not shown in Figure 17) were 61 psi at the front center and 53 psi at the rib. The recorded seal movement during the explosion was >6 in as the seal was destroyed. The two breakwire signals showed that the wires broke as the seal was destroyed near the time of peak pressure loading at ~0.73–0.75 sec.

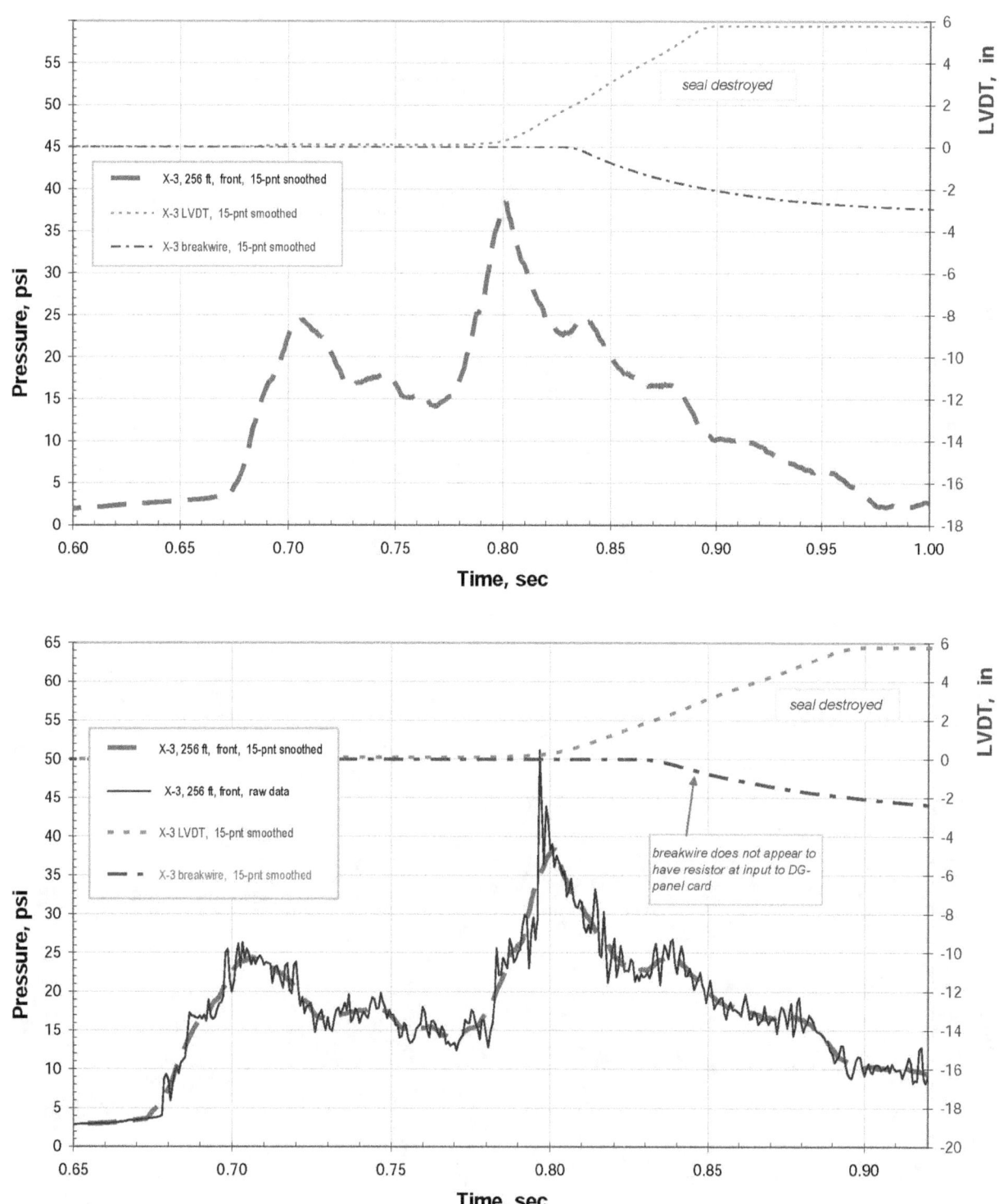

Figure 16.—Pressures and LVDT displacement at Omega block seal in X-3 during Test 2 (LLEM #502).

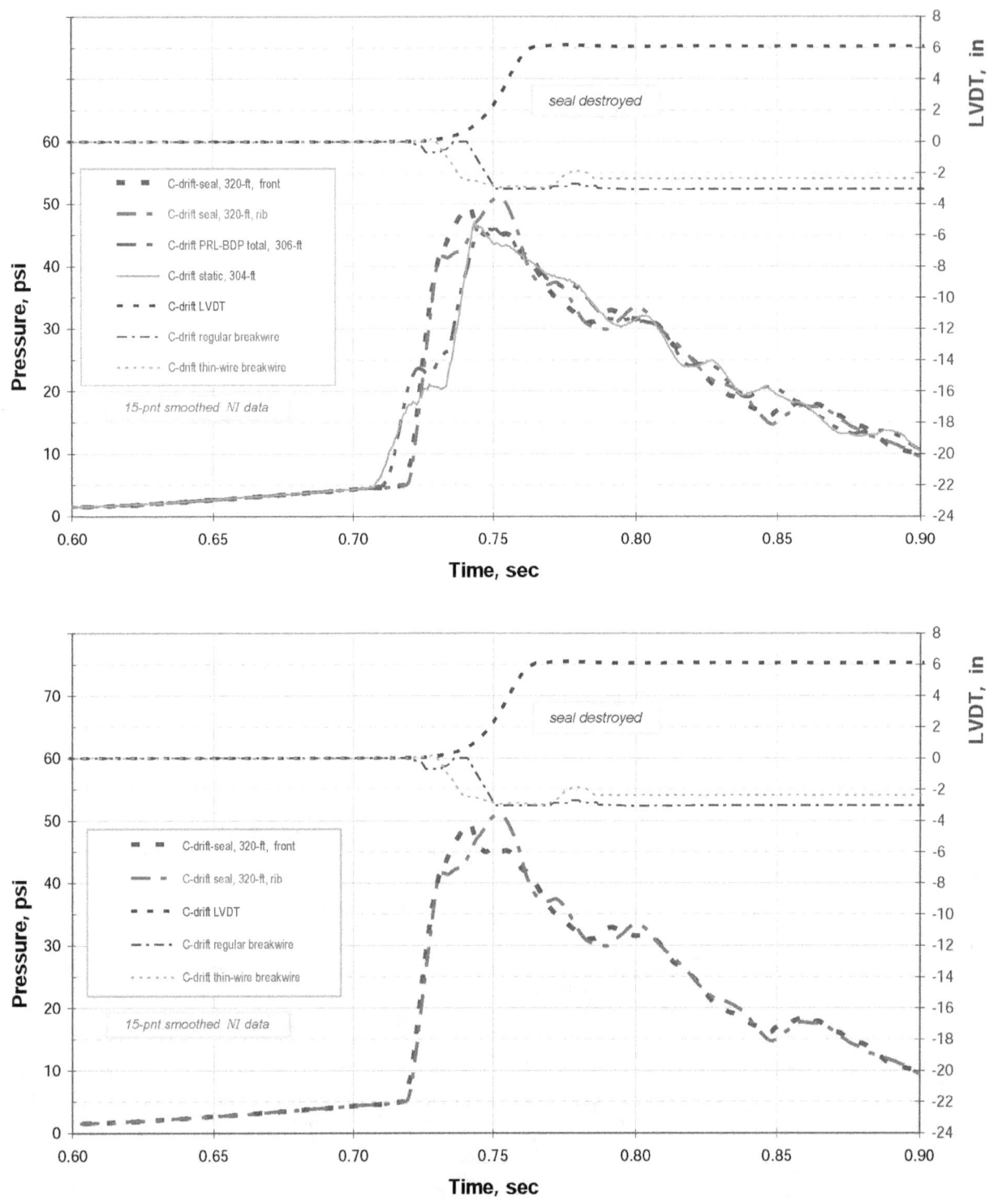

Figure 17.—Pressures and LVDT displacement at C-drift Omega block seal during Test 2 (LLEM #502).

Pressures, LVDT displacement, and breakwire summary data for the seals in X-2, X-3, and C-drift during Test 2 (LLEM #502) are shown in Table 5, similar to the data for Test 1 in Table 2. The pressures at the seals are listed for the NI and KS data acquisition systems. The data were averaged over 10 ms for all of the listed values. The smoothed pressure loading at the middle of the X-2 seal was ~22 psi, and the seal survived the explosion. The smoothed pressure loading at the middle of the X-3 seal was 38.5 psi, and the seal was destroyed. The time that the breakwire on the seal broke was somewhat uncertain due to the slow response time of the breakwire circuit. This was probably caused by a missing resistor. The C-drift seal experienced a smoothed pressure loading of 49.5 psi at the center and 51 psi at the rib. The breakwire data show that the C-drift seal was destroyed at ~0.74 sec.

Table 6 lists quasi-static pressure and flame sensor data at the various DG panels on the walls of B- and C-drifts during LLEM #502. The positions of the Omega block seals in X-2, X-3, and C-drift are depicted by blue shading. On the left of the table are the B-drift wall quasi-static pressures; toward the right are the quasi-static wall pressures in C-drift. The C-drift quasi-static wall pressures remained relatively constant out to ~184 ft, similar to the values for Test 1. The listed pressures increased significantly out to ~320 ft as the pressure pulse was confined by the C-drift seal. The pressure pulse that reflected back toward the face from the seal caused the higher subsequent pressures at 304 ft, at the X-3 seal at 256 ft, and at 234 ft. After the C-drift seal broke, the quasi-static wall pressures beyond the seal were much lower—4.8 psi at 403 ft and 3.9 psi at 501 ft. The pressures decreased even more as the pressure pulse then passed other open crosscuts. The last two columns of Table 6 list the flame signal and arrival time at each of the DG panels. For this test, the flame went past the 184-ft panel but did not reach the 234-ft panel. Therefore, the interpolated flame travel distance was about 210–220 ft. This was the same as for Test 1. Based on the initial gas zone length of 47 ft, the expansion ratio would be about 4.6.

The air leakage data after Test 2 (LLEM #502) are in Table C-4 of Appendix C. The seal in X-2 passed the leakage test. The seals in X-3 and C-drift were destroyed during the test and therefore not measured for air leakage.

Table 5.—Pressures and LVDT displacement at the seals during Test 2 (LLEM #502)

Seal	Seal pressures psi (NI)	Seal pressures psi (KS)	LVDT deflection, in	Break time, sec	Seal		Seal pressures psi (NI)	Seal pressures psi (KS)	LVDT deflection, in	Break time, sec
X-2 BC 156 ft	21.9	22.1	0.025	—						
	Seal survived									
X-3 BC 256 ft	38.5	38.4	>6	~0.83						
	Seal destroyed, debris traveled ~23 ft				C-drift 321 ft	Rib Center	50.9 49.6	50.9 49.4	>6	0.74
					Seal destroyed, debris traveled ~824 ft					

Table 6.—Wall pressures and flame travel during Test 2 (LLEM #502)

B-drift quasi-static pressures				C-drift quasi-static pressures			Flame signal	
Distance, ft	psi (NI)	psi (KS)		Distance, ft	psi (NI)	psi (KS)	Volts	sec (NI)
10	4.5	4.5	1 X-1	13	—	—	>5	0.340
108	3.3	3.3	2	84	24.5	24.2	>5	0.560
158	~3		3 X-2	134	21.7	21.6	>5	0.603
211	2.5	2.5	4	184	23.5	23.1	>5	0.700
257	2.1	2.1	5 X-3	234	30.7	31.1	~0.1	
329	2.8	2.9	6 X-4	304	47.1	46.5	~0	
427	3.0	3.0	7 X-5	403	4.8	4.8	~0	
526	2.9	2.8	8 X-6	501	3.9	3.9	~0	
626	2.9	2.8	9 X-7	598	3.4	3.4	~0	
782	3.0	2.9	10	757	2.8	2.8	~0	
			11	1,506	1.9	1.9		

Figure 18.—Remains of X-3 seal viewed from C-drift after Test 2 (LLEM #502).

Figure 18 shows the remains of the X-3 seal that was destroyed during Test 2 (LLEM #502). There were two very large blocks plus some smaller debris. The debris only traveled ~23 ft toward B-drift from the original seal position in X-3. The long boards in the figure were the header boards for the seal.

The debris from the C-drift seal was thrown much farther than the debris from the X-3 seal. Figure 19 shows the view looking outby from the original seal position in C-drift. On the left is the pressure transducer at the rib. Near the center is the transducer mounted vertically from the roof at the middle of the seal. Just to the right of it is the BDP mounted horizontally on a platform suspended from the roof. At the far right, only a small amount of seal material remains attached to the right rib. Only a few small pieces of seal debris are seen on the floor outby the original seal position.

Figure 20 shows the view looking outby from the DG panel at 403 ft. Even at this distance, the seal debris consists only of small pieces. Larger pieces of debris from the C-drift seal are seen in Figures 21–22, which show the views looking outby from X-6 at 547 ft and X-7 at 647 ft, respectively.

Figure 19.—Postexplosion view looking outby from original seal position in C-drift.

Figure 20.—Postexplosion view looking outby from 403 ft after Test 2 (LLEM #502).

Figure 21.—Debris after Test 2 (LLEM #502) looking outby from X-6 location.

Figure 22.—Debris after Test 2 (LLEM #502) looking outby from X-7 location.

Figure 23.—Debris after Test 2 (LLEM #502) looking outby from 850 ft.

Figure 23 shows the seal debris at a position looking outby at a distance of ~850 ft from the face or ~530 ft from the original C-drift seal position. The farthest the debris traveled during Test 2 was ~824 ft from the original C-drift seal location, as noted in Table 5. There were no wood cribs for this test.

For Test 2 (LLEM #502), the 1,560-lb battery charger was placed at 688 ft from the face of C-drift or 365 ft from the outby face of the C-drift seal. The charger moved ~79 ft during this explosion. Figure 24 shows the final location of the battery charger near 770 ft surrounded by debris from the seal. The total explosion pressure at the 604-ft BDP was 3.8 psi and the dynamic pressure was 0.3 psi. The quasi-static wall pressure at 757 ft was 2.8 psi. The pressure loading at the charger location would have been ~3.5 psi. As for Test 1, the pressure pulse rose to its peak recorded value in less than 7 ms. Therefore, the inby end of the charger would have been subjected to the maximum explosion pressure loading while the outby end was still at ~0 psi. Based on the cross-sectional area of 900 in^2 for the end of the charger and the pressure loading of ~3.5 psi, the total force would have been ~3,100 lb for a few milliseconds. The ~0.3-psi dynamic pressure would have continued to act on the charger for a longer time period. In addition to the air pressure, the battery charger was also hit by debris from the seal, as evidenced in Figure 24.

Figure 24.—Final location of battery charger at ~770 ft.

Some of the roof plates were damaged during the Test 2 explosion, as shown in Figure 25. On the left side of the figure, the plates are shown from inby of the plates, looking outby. On the right side of the figure, the plates are viewed looking across the entry from the DG-panel side. The round plate at 184 ft was slightly bent during the explosion. The round plate at 234 ft from the face was the only one that was severely bent during the explosion. There was little or no obvious damage to the other round plates, square plates, and belt hangers.

Viewed from inby | Viewed across entry from DG panel

184 ft from face

234 ft from face

304 ft from face

Figure 25.—Roof plates at various distances from the face after Test 2 (LLEM #502).

Test 3 (LLEM Test #503), August 4, 2006

The solid-concrete-block seal in X-1 and the 2001-design Omega block seal in X-2 had survived the previous LLEM explosion tests and were left in place for the third test. New Sago nominal 40-in Omega block seals were installed in X-3 and in C-drift at about 320 ft from the face. The intent of Test 3 was to determine if an improperly constructed 40-in-thick Omega block seal (as constructed at the Sago Mine) could meet the regulatory requirement of 20 psi. The new Sago Omega block seals were installed in a manner similar to the previous hybrid Omega block seal in X-3 during Test 1, except the dry layer of BlocBond on the floor was 1.5 in thick (instead of 0.25 in thick) and each course of blocks was positioned completely across the crosscut before applying the properly mixed BlocBond to the top of that course and forcing the BlocBond into the vertical joints using only gloved hands. Each block course was alternated to stagger the block joints. Additional details of the seal construction procedures can be found in Appendix A under "Test No. 3 Protocol" and in Appendix B, sections 4 and 5. Figure 26 shows a schematic of the test setup; the new seals are shown in blue. Wood cribs were also constructed both inby and outby the seal in C-drift, as shown in Figure 26. A hollow-block stopping was constructed outby the C-drift seal at ~384 ft (point D in Figure 26). Another hollow-block stopping was constructed in X-3 between A- and B-drifts (point E in Figure 26). Roof plates that had been damaged in the previous explosion were replaced for this test.

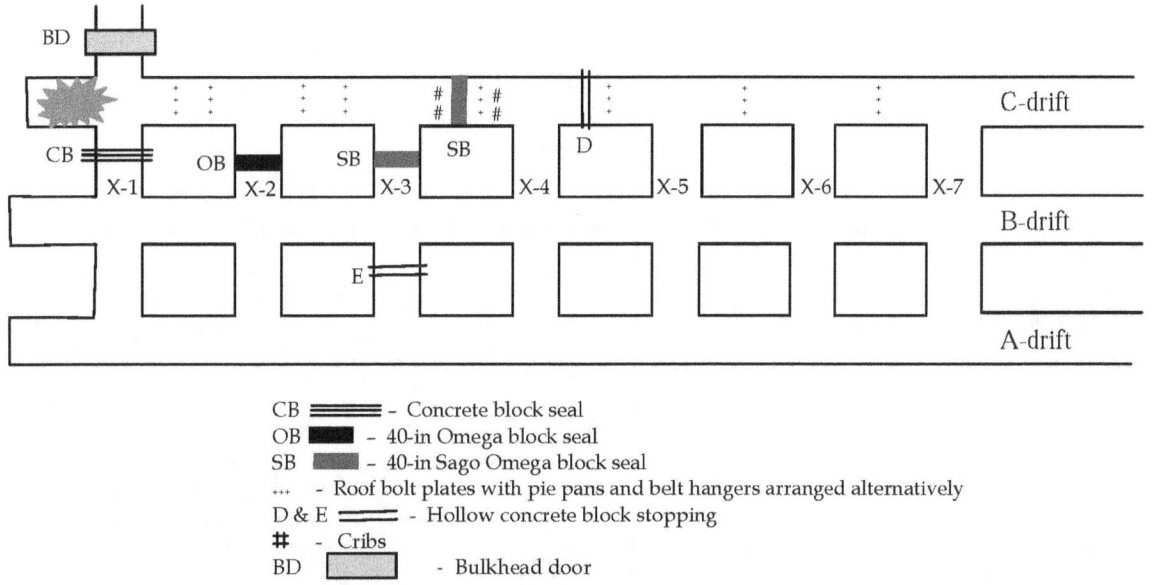

Figure 26.—Setup for Test 3 (LLEM #503).

For Test 3 (LLEM #503), the same 47-ft-long zone of ~10% CH_4 in air was used, but it was ignited at 35 ft from the face of C-drift. This change in the ignition location resulted in a lower pressure explosion propagating out C-drift, as previously observed in the Bruceton Experimental Mine by Nagy [1981, p. 52] and Nagy and Mitchell [1963, p. 20].

In this test, the Omega block seals in X-2, X-3, and C-drift all survived the explosion. Figure 27 shows the smoothed pressure loading at the seal in X-2; Figure 28 shows similar data for the X-3 seal. All of the data were averaged over 10 ms. For this test, because the seals survived, the pressure pulse reflected back and forth between the face and the C-drift seal at 320 ft. This is more obvious in Figure 28 for X-3. The smoothed pressure loading at the X-2 seal was 13.5 psi. At the X-3 seal, there were two pressure transducers—one mounted horizontally and one vertically, as shown in Figure 3. The pressure loading at the X-3 seal was slightly less than 16 psi for both transducers. The maximum recorded LVDT movement during the explosion was 0.01 in for the X-2 seal and 0.04 in for the X-3 seal. Since the seals did not fail, there were no changes in the breakwire signals.

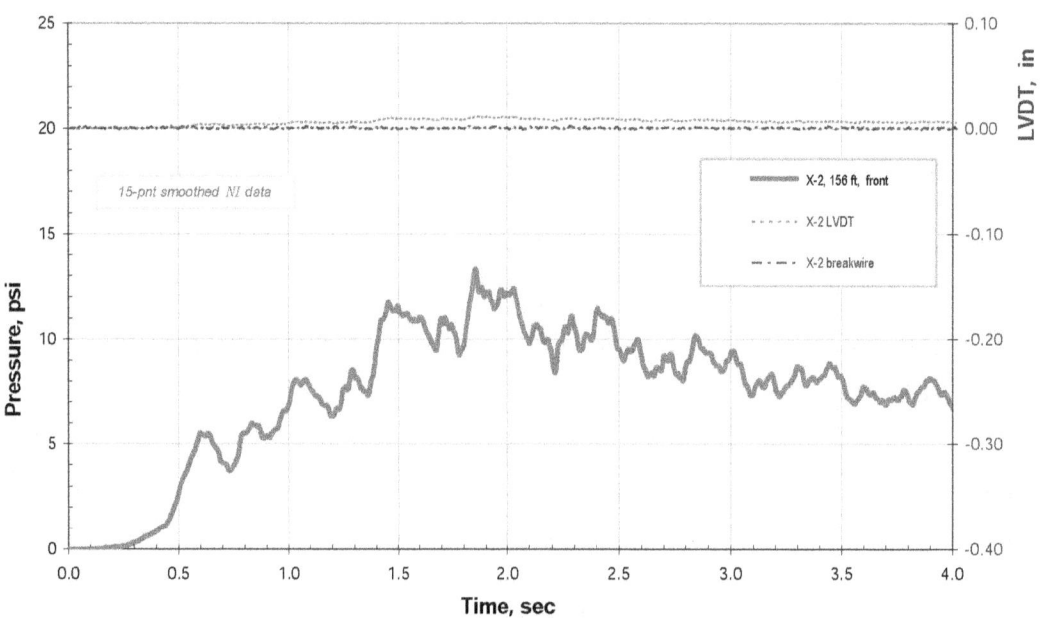

Figure 27.—Pressures and LVDT displacement at the X-2 seal during Test 3 (LLEM #503).

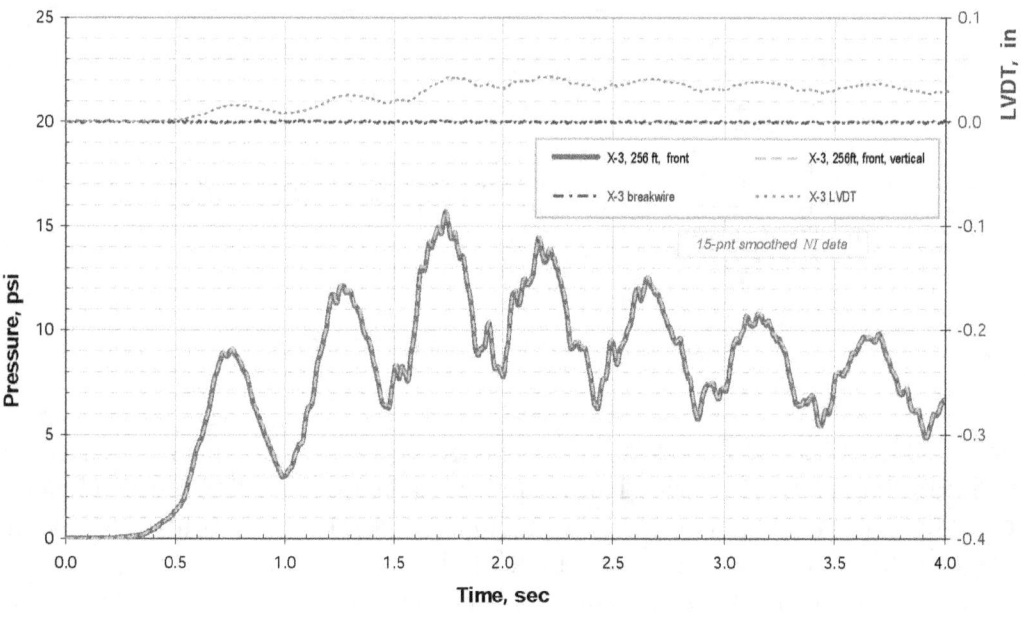

Figure 28.—Pressures and LVDT displacement at the X-3 seal during Test 3 (LLEM #503).

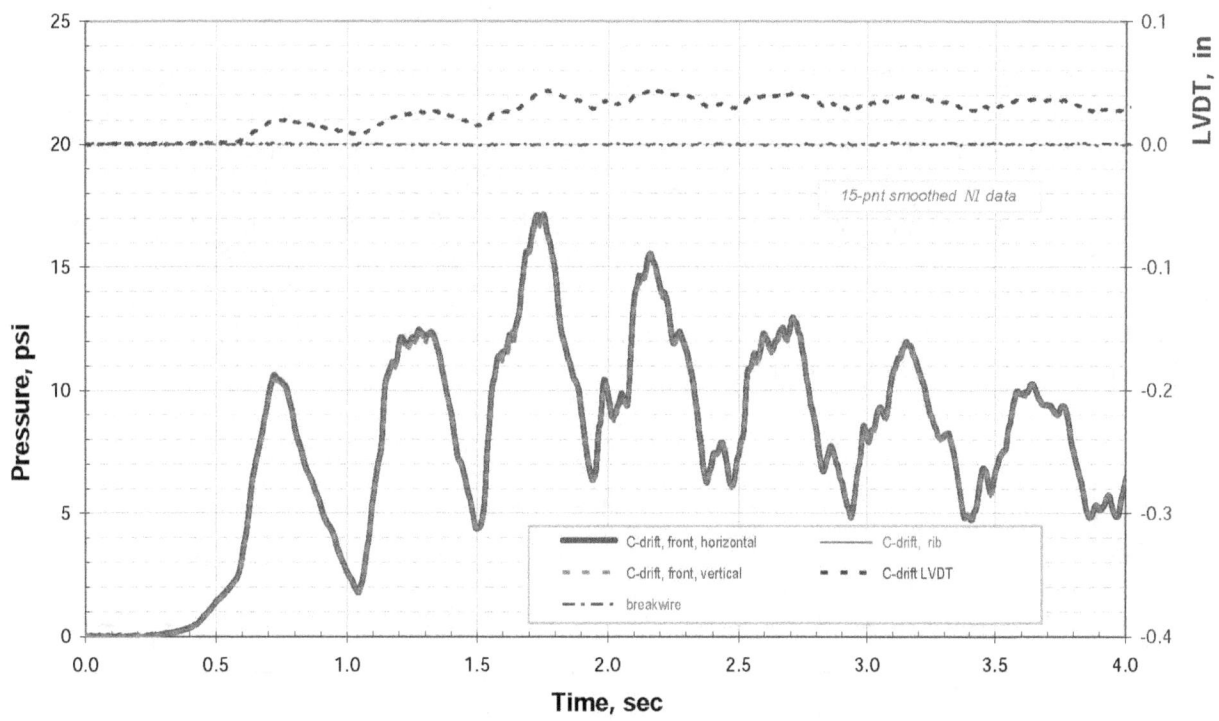

Figure 29.—Pressures and LVDT displacement at the C-drift seal during Test 3 (LLEM #503).

Figure 29 shows data from three pressure transducers at the C-drift seal at 320 ft from the face. Two transducers were near the middle of the seal—one mounted horizontally and one vertically, as shown in Figure 3. The third transducer was at the rib, the same as for the previous test. The pressure at all three transducers was essentially the same as shown in the figure. The smoothed pressure loadings at the horizontal and vertical pressure transducers were 17.2 psi and 17.0 psi, respectively. Since there was little difference in the two values, this shows that the wind or dynamic pressure was approximately zero at the seal. This would be expected since the wind velocity has to go to zero as the pressure wave reaches the seal. The smoothed pressure loading at the rib was 17.2 psi. The LVDT data in the upper part of the figure show that the seal moved about 0.04 in during the explosion.

The air leakage data after Test 3 (LLEM #503) are in Table C-6 of Appendix C. All three seals passed the leakage test. In addition to the three seals that survived the explosion, the wood cribs and stoppings also survived. There was no obvious damage to the roof plates and belt hangers.

For this test, the flame went past the 134-ft DG panel, but did not reach the 184-ft DG panel. Therefore, the interpolated flame travel distance was about 160 ft. This was a shorter distance than for the first two tests even though the initial gas zone length was 47 ft for all three tests. The reason was the change in ignition location, which led to slower burning of the gas and more heat losses.

Test 4 (LLEM Test #504), August 16, 2006

All three seals, the wood cribs, and the hollow-block stoppings had survived the previous LLEM explosion test and were left in place for the fourth test. Since the peak explosion pressures in Test 3 were under 20 psi, the intent of Test 4 was to increase the pressure slightly above Test 3 to determine if an improperly constructed 40-in-thick Omega block seal (as constructed at the Sago Mine) could meet the regulatory requirement of 20 psi. Figure 30 shows a schematic of the test setup, which was almost the same as for the previous test. For Test 4, the same 47-ft-long zone of ~10% CH_4 in air was used, but the ignition point was moved from 35 to 28.5 ft from the face of C-drift to achieve slightly higher pressures than for Test 3. Additional details of the test procedure can be found in Appendix A under "Test No. 4 Protocol."

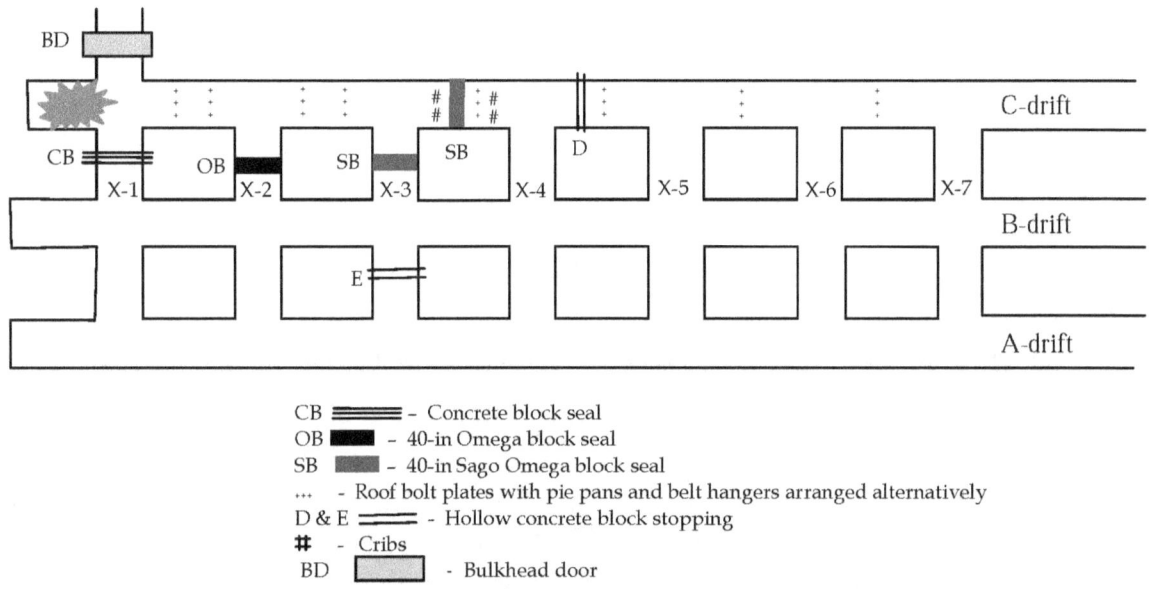

Figure 30.—Setup for Test 4 (LLEM #504).

In Test 4 (LLEM #504), the Omega block seals in X-2, X-3, and C-drift all survived the explosion. Figure 31 shows the smoothed pressure loading at the seal in X-2; Figure 32 shows similar data for the X-3 seal. Because the seals survived, the pressure pulse reflected back and forth between the face and the C-drift seal at 320 ft. This is more obvious in Figure 32 for X-3. The smoothed pressure loading at the X-2 seal was 15 psi. At the X-3 seal, there were two pressure transducers—one mounted horizontally and one vertically, as shown in Figure 3. The pressure loading at the X-3 seal was slightly greater than 18 psi for both transducers. The maximum recorded LVDT movement during the explosion was 0.01 in for the X-2 seal and 0.09 in for the X-3 seal. Since the seals did not fail, there were no changes in the breakwire signals.

Figure 33 shows data from three pressure transducers at the C-drift seal at 320 ft from the face. As for the previous test, two transducers were near the middle of the seal—one mounted horizontally and one vertically, as shown in Figure 3. The third transducer was at the rib. The pressure loading at all three transducers was essentially the same as shown in Figure 33. The smoothed pressure loadings at the horizontal and vertical pressure transducers were 20.6 psi and 20.4 psi, respectively. This shows that the dynamic pressure loading was approximately zero at the seal. This would be expected since the wind velocity has to go to zero as the pressure wave reaches the seal. The smoothed pressure loading at the rib was 20.5 psi. Figure 33 also shows the pressure pulse reflecting back and forth between the face and the C-drift seal, similar to Figure 32. The LVDT data in the upper part of the figure show that the seal moved about 0.07 in during the explosion.

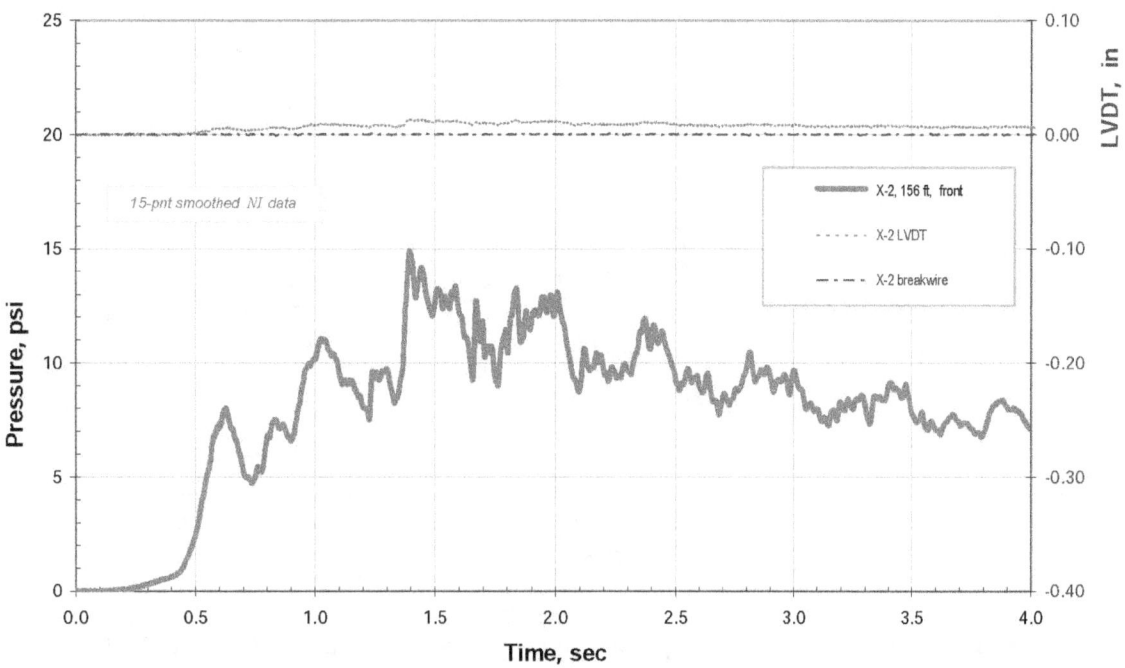

Figure 31.—Pressures and LVDT displacement at the X-2 seal during Test 4 (LLEM #504).

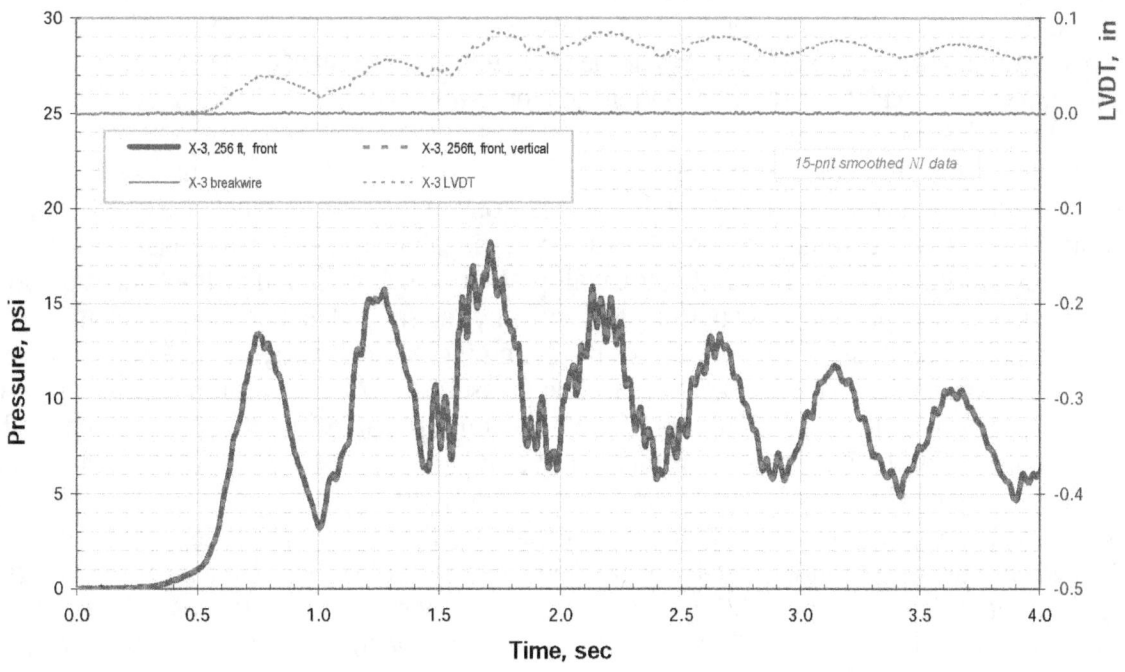

Figure 32.—Pressures and LVDT displacement at the X-3 seal during Test 4 (LLEM #504).

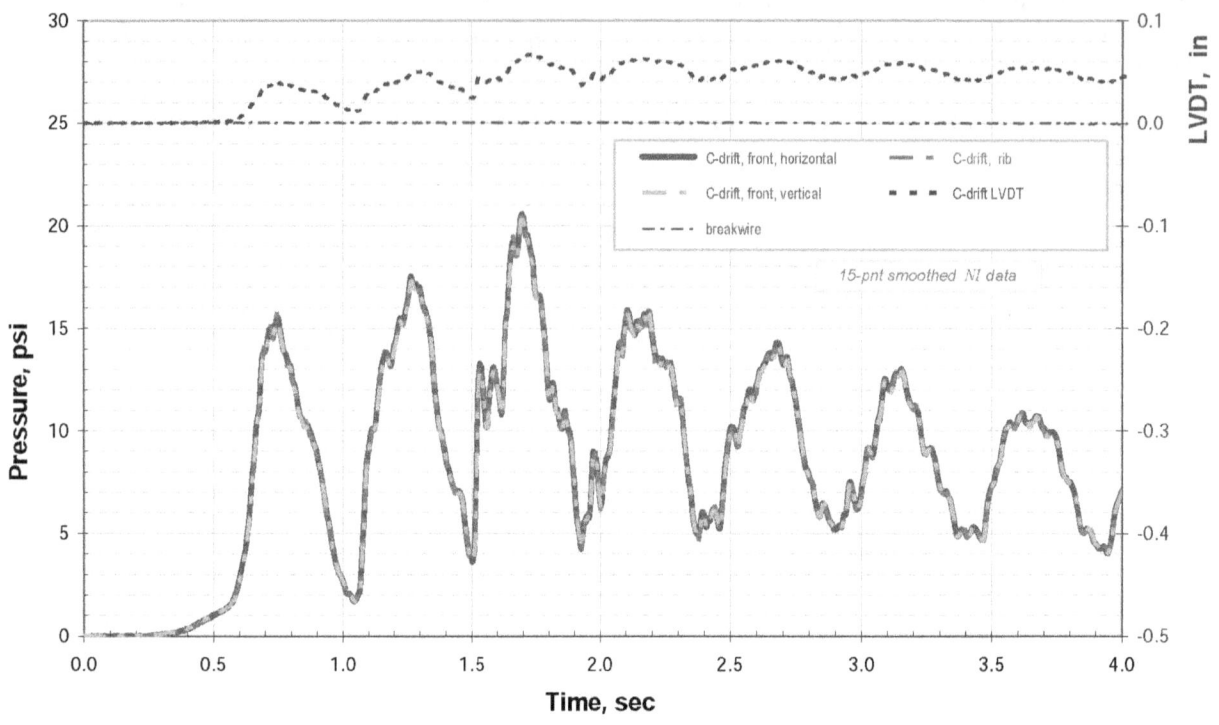

Figure 33.—Pressures and LVDT displacement at the C-drift seal during Test 4 (LLEM #504).

Pressures, LVDT displacement, and breakwire summary data for the seals in X-2, X-3, and C-drift during Test 4 (LLEM #504) are shown in Table 7. The pressures at the seals are listed for the NI and KS data acquisition systems. The data were averaged over 10 ms for all of the listed pressures. The smoothed pressure loadings at the middle of the X-2 and X-3 seals were ~15 psi and ~18 psi, respectively. The C-drift seal experienced a pressure loading of ~20.5 psi. All three seals survived the explosion.

Table 8 lists quasi-static wall pressure and flame sensor data at the various DG panels on the walls of B- and C-drifts during LLEM #504. The positions of the Omega block seals in X-2, X-3, and C-drift are depicted by blue shading. On the left of the table are the B-drift quasi-static wall pressures; toward the right are the quasi-static wall pressures in C-drift. All of the B-drift pressures were ~0 because the explosion was confined to C-drift by the seals. The C-drift quasi-static wall pressures remained relatively constant out to ~184 ft and increased slightly out to ~320 ft as the pressure pulse was confined by the C-drift seal. The pressures beyond the seal in C-drift were ~0 since the seals survived the explosion. The last two columns of Table 8 list the flame signal and arrival time at each of the DG panels. For this test, the flame went past the 134-ft DG panel but did not reach the 184-ft DG panel. Therefore, the interpolated flame travel distance was about 160 ft. This was a distance similar to that in Test 3, which also had the ignition point outby the face.

The air leakage data after Test 4 (LLEM #504) are in Table C-7 of Appendix C. All three seals passed the leakage test. In addition to the three seals that survived the explosion, the wood cribs and stoppings also survived. There was no obvious damage to the roof plates and belt hangers.

Table 7.—Pressures and LVDT displacement at the seals during Test 4 (LLEM #504)

Seal		Seal pressures		LVDT deflection, in	Break time, sec	Seal		Seal pressures		LVDT deflection, in	Break time, sec
		psi (NI)	psi (KS)					psi (NI)	psi (KS)		
X-2 BC 156 ft	H	14.9	14.9	0.01	—						
		Seal survived									
X-3 BC 256 ft	H V	18.2 18.2	18.2 18.2	0.09	—						
		Seal survived									
						C-drift 321 ft	Rib H V	20.5 20.6 20.4	20.5 20.6 20.4	0.07	—
								Seal survived			

H Horizontal. V Vertical.

Table 8.—Wall pressures and flame travel during Test 4 (LLEM #504)

B-drift quasi-static pressures				C-drift quasi-static pressures			Flame signal	
Distance, ft	psi (NI)	psi (KS)		Distance, ft	psi (NI)	psi (KS)	Volts	sec (NI)
10	0.0	0.0	1	13	—	—	>5	0.807
			X-1					
108	0.0	0.0	2	84	16.6	16.6	>5	0.448
158	~0	~0	3	134	15.3	15.3	>5	0.621
			X-2					
211	~0	~0	4	184	15.3	15.3	~0	
257	0.0	0.0	5	234	17.2	17.1	~0	
			X-3					
329	0.0	0.0	6	304	19.9	19.8	~0	
			X-4					
427	0.0	0.0	7	403	0.0	0.0	~0	
			X-5					
526	0.0	0.0	8	501	0.0	0.0	~0	
			X-6					
626	0.0	0.0	9	598	0.0	0.0	~0	
			X-7					
782	0.0	0.0	10	757	0.0	0.0	~0	
			11	1,506	0.0	0.0		

Test 5 (LLEM Test #505), August 23, 2006

All three seals, the wood cribs, and the hollow-block stoppings had survived the previous two LLEM explosion tests and were left in place for the fifth test. Figure 34 shows a schematic of the test setup, which was almost the same as for the previous test. In addition to the seals, there were hollow-block stoppings at 384 ft in C-drift and in X-3 between A- and B-drifts. There were also wood cribs both inby and outby the C-drift seal, as shown in Figure 34. For Test 5, the same 47-ft-long zone of ~10% CH_4 in air was used, but the ignition point was moved back to the face of C-drift to achieve pressures similar to those in Test 2. Additional details of the test procedure can be found in Appendix A under "Test No. 5 Protocol."

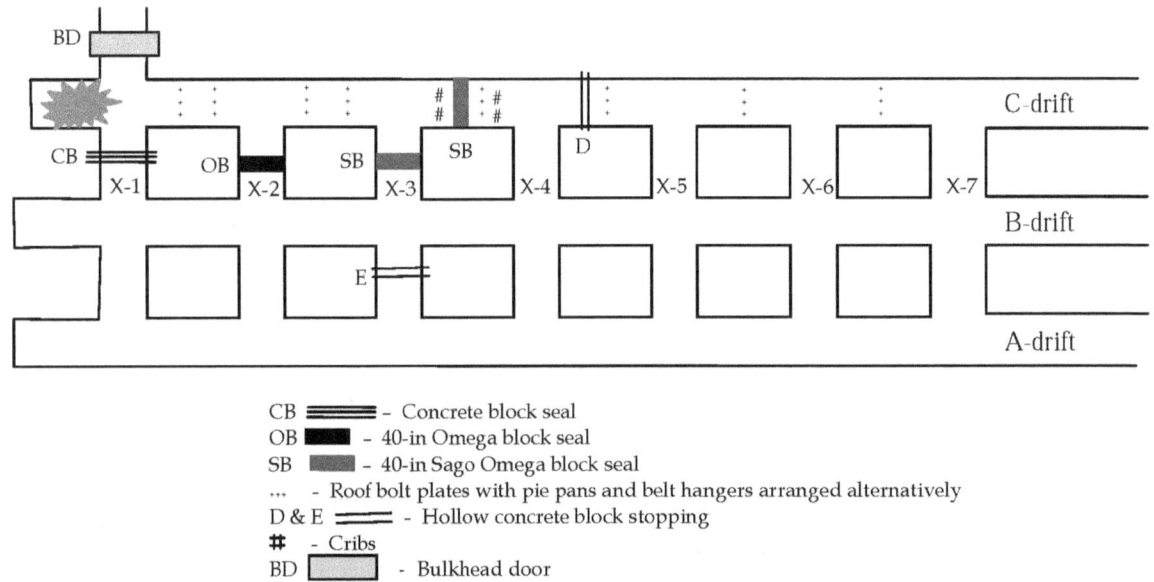

Figure 34.—Setup for Test 5 (LLEM #505).

For Test 5 (LLEM #505), the 47-ft-long ignition zone at the face was filled with 661 ft^3 of natural gas to give a mixture of ~10% CH_4 in air, the same as for Tests 1 through 4. This CH_4-air zone was ignited at the face of C-drift, and the pressure pulse propagated out C-drift past the seals in X-2 and X-3 to the seal in C-drift. Because the gas was ignited at the face, the resulting explosion pressures were much higher than those in Tests 3 and 4.

In this test, the Omega block seal in X-2 survived, but the Omega block seals in X-3 and C-drift were destroyed by the higher pressures during Test 5. Figure 35 shows the pressure loading at the seal in X-2 along with the quasi-static pressure at the wall of C-drift inby the seal at 134 ft and outby the seal at 184 ft. The pressure transducer at the seal was located approximately in the middle front of the seal at 156 ft from the face of C-drift. Most of the data are from the NI data acquisition system. The pressure data at the X-2 seal are from the KS data acquisition system because there was a problem with that channel on the NI system during this test. All of the data were averaged over 10 ms. The pressure loading at the X-2 seal was 26 psi. The maximum recorded LVDT movement during the explosion was 0.03 in. Since the seal did not fail, there was no change in the breakwire signal.

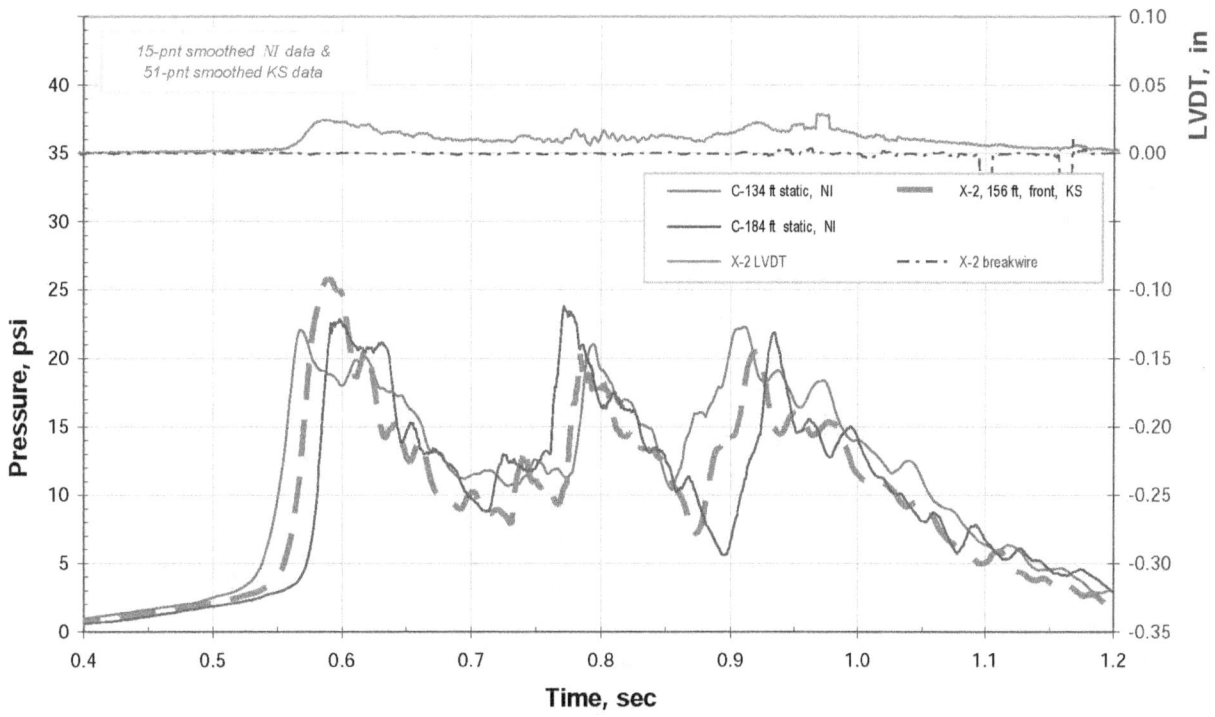

Figure 35.—Pressures and LVDT displacement at the X-2 seal during Test 5 (LLEM #505).

Figure 36 shows the pressure loading and LVDT displacement data at the X-3 seal, along with the quasi-static pressure at the wall of C-drift inby the seal at 234 ft. The pressure transducers at the seal were located near the middle front of the seal at 256 ft from the face of C-drift. There were two pressure transducers at the seal—one mounted horizontally and one vertically, as shown in Figure 3. The LVDT and breakwire data at the top of the figure show that the X-3 seal survived the initial outgoing pressure loading of ~29 psi at ~0.65 sec. The seal was destroyed by the subsequent higher pressure loading (~33–35 psi) that was reflected back from the C-drift seal at ~0.74 sec.

Figure 37 shows the pressures from the NI data acquisition system at the X-3 seal on an expanded time scale. Both the raw data at 1,500 samples per second and the smoothed data (15-point smoothing or 10-ms average) are shown. The averaged pressure loading for the horizontal transducer was 34.7 psi. The peak pressure loading from the raw data for the horizontal transducer at the X-3 seal was ~42 psi, but it lasted only about 1 ms. The averaged pressure loading for the vertical transducer at the X-3 seal was 32.8 psi. The peak pressure loading from the raw data for this vertical transducer at the X-3 seal was ~38 psi, but it lasted only about 1 ms. The curve at the top of the figure shows the seal displacement as measured by the LVDT on the back of the seal. The maximum recorded movement during the explosion was >6 in as the seal was destroyed. The breakwire raw data at the top of Figure 37 are shown as the dotted curves that almost instantaneously drop below the baseline; the units are arbitrary. The breakwire signals showed that the wires broke at 0.743–0.753 sec after ignition.

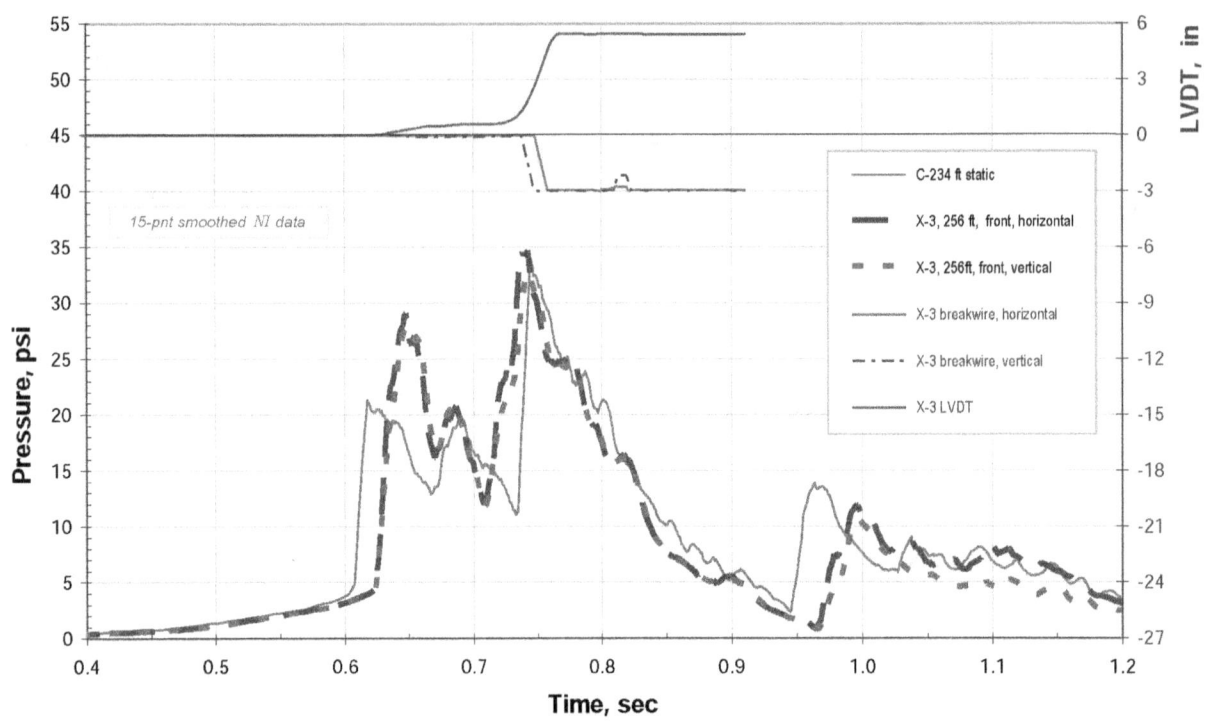

Figure 36.—Pressures and LVDT displacement at the X-3 seal during Test 5 (LLEM #505).

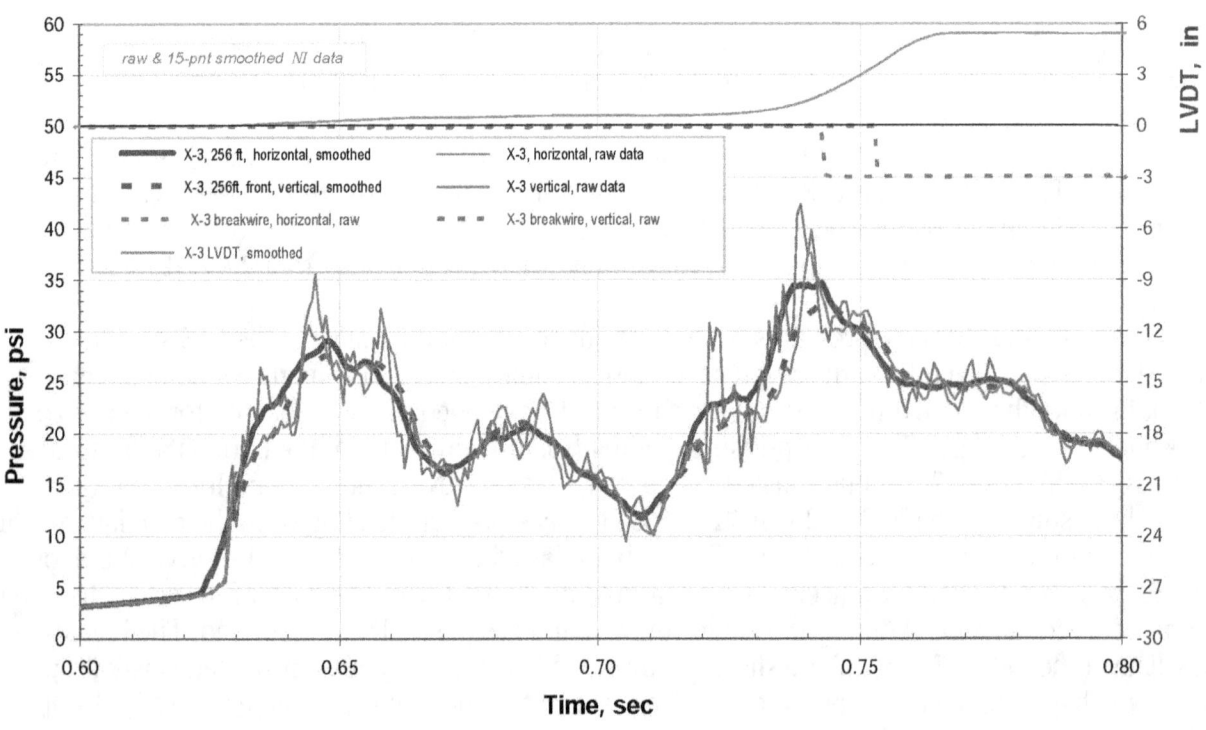

Figure 37.—Pressures and LVDT displacement on an expanded time scale for the X-3 seal during Test 5 (LLEM #505).

Figure 38 shows NI data from three pressure transducers at the C-drift seal at 320 ft from the face. Two transducers were near the middle of the seal—one mounted horizontally and one vertically, as shown in Figure 3. A third transducer was mounted on the rib perpendicular to the entry. The total measured explosion pressure loadings at all three transducers were similar, as shown in the figure. The LVDT and breakwire data show that the seal broke near the time of peak pressure loading.

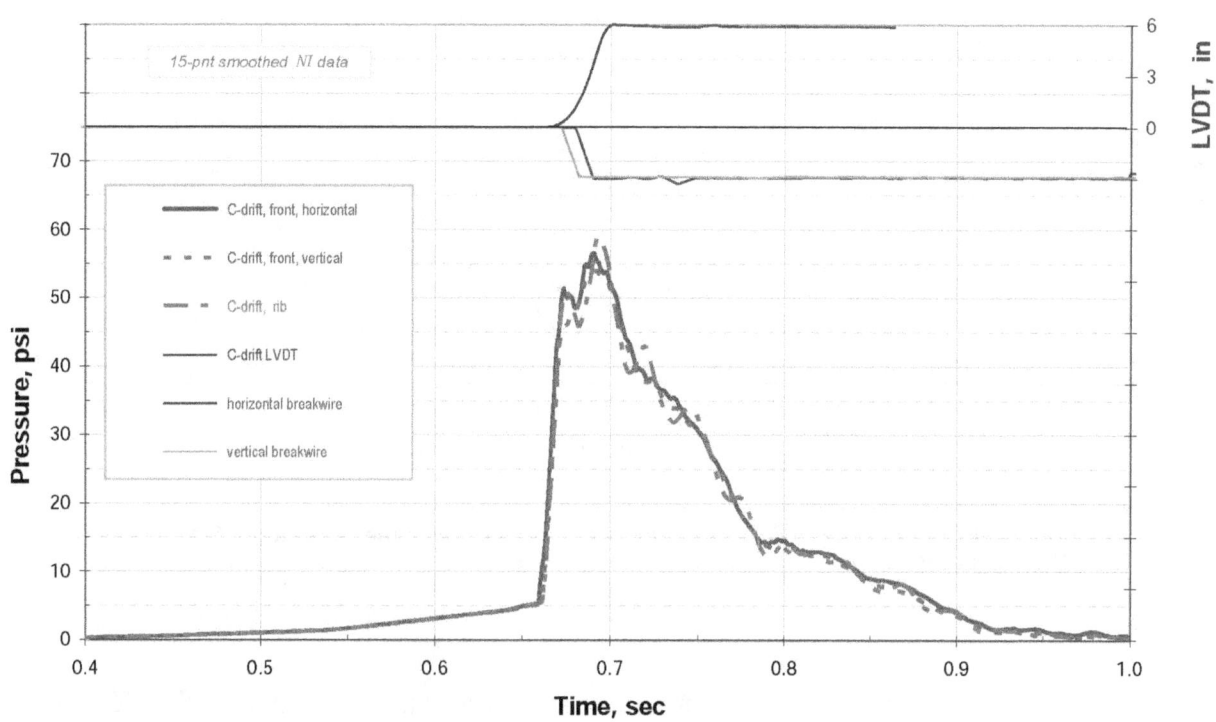

Figure 38.—Pressures and LVDT displacement at the C-drift seal during Test 5 (LLEM #505).

Figure 39 shows the pressures from the NI data acquisition system at the seal on an expanded time scale. Both the raw data at 1,500 samples per second and the smoothed data (15-point smoothing or 10-ms average) are shown. The smoothed pressure loading for the horizontal transducer was 56.6 psi. The peak pressure loading from the raw data for the horizontal transducer was ~63 psi, but it lasted only about 1 ms. The smoothed pressure for the vertical transducer was 54.1 psi. The peak pressure loading from the raw data for the vertical transducer was ~57 psi, but it lasted only about 1 ms. The smoothed pressure loading at the rib was 59.1 psi, and the peak value from the raw data was ~63 psi. The LVDT data in the upper part of the figure show that the seal moved >6 in as the seal was destroyed. The two breakwire signals (from the NI raw data) both show a sharp discontinuity as the wires broke at 0.677–0.685 sec.

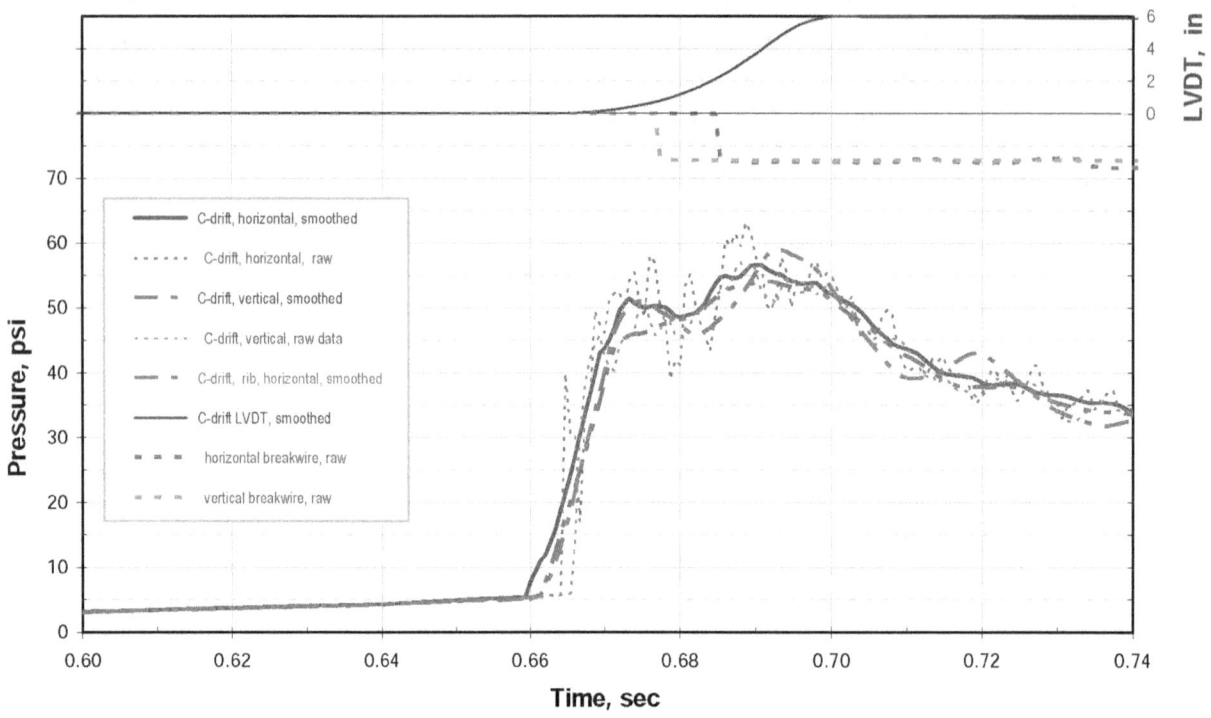

Figure 39.—Pressures and LVDT displacement on an expanded time scale at the C-drift seal during Test 5 (LLEM #505).

Pressure, LVDT displacement, and breakwire summary data for the seals in X-2, X-3, and C-drift during Test 5 (LLEM #505) are shown in Table 9. The explosion pressure loadings at the seals are listed for the NI and KS data acquisition systems. The data were averaged over 10 ms for the listed pressure values. The breakwire time is from the NI raw data and is the time after ignition. The smoothed pressure loading at the middle of the X-2 seal was 25.8 psi (from the KS data acquisition system), and the seal survived the explosion. The smoothed pressure loading from the horizontal transducer at the middle of the X-3 seal was 34.7 psi and 33.1 psi from the NI and KS data acquisition systems, respectively. The smoothed pressure loading from the vertical transducer at the middle of the X-3 seal was 32.8 psi from both systems. The X-3 seal breakwire raw data show that the vertical wire broke at 0.743 sec and the horizontal wire broke at 0.753 sec as the seal was destroyed. The pressure loading in front of the stopping at point E (Figure 34) was 4.0 psi, and the stopping was partially destroyed. The stopping breakwire did not break until 2.17 sec.

The C-drift seal experienced a smoothed pressure loading from the horizontal transducer (middle front of seal) of 56.6 psi and 55.7 psi from the NI and KS data acquisition systems, respectively. The smoothed pressure loading from the vertical transducer (middle front of seal) was 54.1 psi from both systems. The smoothed pressure loading at the rib was 59.1 psi and 58.9 psi from the NI and KS systems, respectively. The breakwire raw data show that the vertical wire broke at 0.677 sec and the horizontal wire broke at 0.685 sec as the seal was destroyed. The breakwire data show that the C-drift seal was destroyed before the X-3 seal. The smoothed pressure loading in front of the stopping in C-drift at 384 ft was 12.4 psi, and the stopping was totally destroyed. The breakwire raw data show that the stopping at point D (Figure 34) was destroyed at 1.092 sec.

Table 9.—Pressures and LVDT displacement at the seals during Test 5 (LLEM #505)

Seal or stopping		Seal pressures psi (NI)	Seal pressures psi (KS)	LVDT deflection, in	Break time, sec	Seal or stopping		Seal pressures psi (NI)	Seal pressures psi (KS)	LVDT deflection, in	Break time, sec
X-2 BC 156 ft Seal	H	—	25.8	0.03	—						
		Seal survived									
X-3 BC 256 ft Seal	H	34.7	33.1	>6	0.753						
	V	32.8	32.8		0.743	Wood cribs	A				0.671
							B				~0.665
		Seal destroyed, debris traveled >108 ft to far wall of A-drift									
X-3 AB Stopping	H	4.0	4.0		2.17	C-drift 320 ft Seal	Rib	59.1	58.9		
							H	56.6	55.7	>6	0.685
		Stopping partially destroyed, debris traveled to far wall of A-drift					V	54.1	54.1		0.677
						Seal destroyed, debris traveled ~556 ft					
						C-drift 384 ft Stopping	H	12.4	12.4		1.092
						Stopping destroyed, debris traveled ~460 ft					

H Horizontal. V Vertical.

Table 10 lists quasi-static pressure and flame sensor data at the various DG panels on the walls of B- and C-drifts during LLEM #505. The positions of the Omega block seals in X-2, X-3, and C-drift are depicted by blue shading. On the left of the table are the B-drift quasi-static wall pressures; toward the right are the quasi-static wall pressures in C-drift. The C-drift pressures remained relatively constant out to ~184 ft, similar to the values in Test 2. The quasi-static wall pressure increased significantly out to ~320 ft as the pressure pulse was confined by the C-drift seal. The pressure pulse that reflected back toward the face from the seal caused higher pressures at 304 ft, the X-3 seal at 256 ft, and at 234 ft. After the seal broke, the explosion pressures beyond the seal were much lower: 12.4 psi at the stopping at 384 ft, ~3.6 psi at 403 ft, and ~3 psi at 501 ft, etc. The last two columns of Table 10 list the flame signal and arrival time at each of the DG panels. For this test, the flame went past the 184-ft panel, but did not reach the 234-ft panel. Therefore, the interpolated flame travel distance was about 210 ft, similar to the travel distances for Tests 1 and 2. Based on the initial gas zone length of 47 ft, the expansion ratio would be about 4.5.

Table 10.—Wall pressures and flame travel during Test 5 (LLEM #505)

B-drift quasi-static pressures				C-drift quasi-static pressures			Flame signal	
Distance, ft	psi (NI)	psi (KS)		Distance, ft	psi (NI)	psi (KS)	Volts	sec (NI)
				3	27.0	27.1		
10	4.6	—	1	13	—	—	>5	0.298
			X-1					
108	3.4	—	2	84	25.4	24.7	>5	0.513
158	4.2	—	3	134	22.3	22.3	>5	0.543
			X-2					
211	3.7	—	4	184	23.8	23.1	>5	0.643
257	2.4	—	5	234	33.9	33.2	~0	
			X-3					
329	4.0	—	6	304	55.1	54.0	~0	
			X-4					
427	3.1	—	7	403	3.7	3.5	~0	
			X-5					
526	2.6	—	8	501	3.1	2.8	~0	
			X-6					
626	2.9	—	9	598	2.6	2.5	~0	
			X-7					
782	2.8	—	10	757	2.6	2.7	~0	
			11	1,506	1.6	1.6		

Figure 40 shows the seal in X-2 that survived Test 5 (LLEM #505). The air leakage data are in Table C-8 in Appendix C. The X-2 seal passed the leakage test. The seals in X-3 and C-drift were destroyed during the test and therefore not measured for air leakage.

Figures 41–42 show the remains of the X-3 seal that was destroyed during Test 5 (LLEM #505). Figure 41 shows the view looking toward the original X-3 seal position from slightly outby in C-drift. Very little debris can be seen in C-drift. The vertical post holding the pressure transducers is still intact. The original seal position was just behind the pressure transducer post. The long board hanging from the roof in the figure was one of the header boards for the seal. Figure 42 shows the view looking directly into X-3 from C-drift. A large amount of debris can be seen beyond the original seal location and extending into the B-drift intersection. Figure 43 shows additional debris from the X-3 seal piled against the stopping between A- and B-drifts. There is also debris beyond the stopping into A-drift. The left side of the stopping was mostly destroyed, but the right side remained intact. The pressure transducer post in front of the stopping also survived. Figure 44 shows the debris from the X-3 seal and stopping piled against the far wall of A-drift.

Figure 40.—Seal in X-2 that survived Test 5 (LLEM #505).

Figure 41.—Debris from X-3 seal after Test 5 (LLEM #505) viewed from C-drift.

Figure 42.—Debris from X-3 seal after Test 5 (LLEM #505) looking into X-3 toward B-drift.

Figure 43.—Debris from the X-3 seal at the X-3 stopping location after Test 5 (LLEM #505).

Figure 44.—Debris from the X-3 seal and stopping piled against the far wall of A-drift after Test 5 (LLEM #505).

The debris from the C-drift seal was thrown much farther than the debris from the X-3 seal, which was stopped by the A-drift wall. Figure 45 shows the postexplosion view looking outby near the original seal position in C-drift. Near the center is the pressure transducer post. The original seal location was just behind this post. Only a few wood blocks and small pieces of seal debris are seen on the floor outby the original seal position.

Figure 46 shows the view looking outby toward X-4 at 355 ft. Wood blocks and small pieces of seal debris are seen in the photo. Some debris was carried into X-4 by the explosion. Figure 47 shows the view looking outby from the DG panel at 403 ft. At this distance, large amounts of wood and seal debris were found. Figure 48, looking outby from X-6 at 547 ft, also shows large amounts of debris from the wood cribs, C-drift seal, and C-drift stopping.

Figure 45.—Postexplosion view in C-drift looking outby toward original seal location at 320 ft.

Figure 46.—Debris in C-drift after Test 5 (LLEM #505) looking outby toward X-4 (right side of photo) at 355 ft.

Figure 47.—Debris from C-drift seal, stopping, and wood cribs looking outby from 403 ft.

Figure 48.—Debris in C-drift after Test 5 (LLEM #505) looking outby from X-6 at 547 ft.

Figure 49 shows the seal debris at a position looking outby from a distance of ~600 ft from the face or ~280 ft from the original seal location. In the distance is X-7 at 647 ft. Pieces of Omega blocks from the seal and pieces of hollow concrete blocks from the stopping can be seen in the photo.

Figure 49.—Debris in C-drift after Test 5 (LLEM #505) looking outby from ~600 ft.

Figure 50 shows the seal debris looking outby in C-drift from the DG panel at 757 ft from the face or ~435 ft from the original seal location. There are numerous large pieces of seal debris in the photo. Figure 51 shows the debris at ~850 ft after Test 5. There are still some large pieces of seal debris at this location. In the upper right of the photo is a carbon monoxide sensor box hanging from the roof. There was no apparent damage to the sensor box during the explosion.

Figure 50.—Debris in C-drift after Test 5 (LLEM #505) looking outby from 757 ft.

Figure 51.—Debris in C-drift after Test 5 (LLEM #505) at ~850 ft.

Figure 52 shows the last piece of seal debris at ~880 ft. During Test 5 (LLEM #505), pieces of Omega blocks traveled as far as 556 ft from the original C-drift seal location, as noted in Table 9 and shown in Figure 52. Debris from the C-drift stopping traveled ~460 ft from the original stopping location. Debris from the wood cribs traveled ~437 ft. These values were measured from the postexplosion surveys.

Figure 52.—Debris in C-drift after Test 5 (LLEM #505) at ~880 ft.

For Test 5 (LLEM #505), the 1,560-lb battery charger was placed at 688 ft from the face of C-drift or 365 ft from the outby face of the C-drift seal. Figure 53 shows the final location of the battery charger after the test, with large amounts of seal debris. The charger moved 30 ft during this explosion. The total explosion pressure at the 604-ft BDP was 2.7 psi, and the dynamic pressure was 0.2 psi. The quasi-static wall pressure at 757 ft was 2.6 psi. The pressure loading at the charger would have been ~2.5 psi. However, for Test 5 the pressure pulse took much longer than 7 ms to reach its maximum value. Therefore, for this test, the maximum differential pressure loading from the inby to outby end of the charger was only ~0.33 psi. The cross-sectional area of 900 in^2 and differential explosion pressure loading of ~0.33 psi would result in a total force of ~300 lb for a few milliseconds. The 0.2-psi dynamic pressure (~180-lb force) would have continued to act on the charger for a longer period of time. In addition to the air pressure, the battery charger was also hit by debris from the seal, stopping, and wood cribs, as evidenced in Figure 53.

Figure 53.—Final location of battery charger near 720 ft after Test 5 (LLEM #505).

Some of the roof plates were damaged during the Test 5 explosion, as shown in Figures 54–55. On the left side of the figures, the plates are shown from inby of the plates, looking outby. On the right side of the figure, the plates are viewed looking across the entry from the DG-panel side. There was only a slight bending of the round plate at 84 ft. The round plates at 134, 184, and 234 ft from the face were severely bent during the explosion. At 304 ft, the round plate was slightly bent, but there was no obvious damage to the square plate. There was only some minor impact damage to the round plate at 403 ft, but no obvious damage to the square plate. However, the belt hanger at 403 ft was significantly bent, as shown in Figures 56–57. These views are looking across the entry toward the DG panel; therefore, outby is to the right in these photos. This belt hanger was probably damaged by flying debris because the inby belt hangers were not damaged even though they were exposed to greater explosion pressures. There was little or no obvious damage to the other round plates, square plates, and belt hangers throughout C-drift.

Viewed from inby | Viewed across entry from DG panel

84 ft from face

134 ft from face

184 ft from face

Figure 54.—Roof plates at 84, 134, and 184 ft from the face after Test 5 (LLEM #505).

Viewed from inby   Viewed across entry from DG panel

234 ft from face

304 ft from face

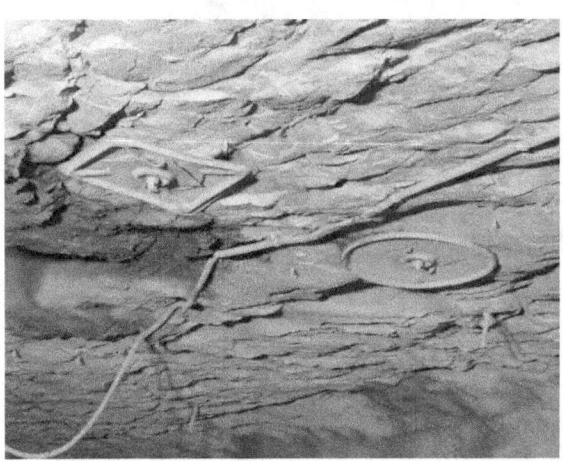 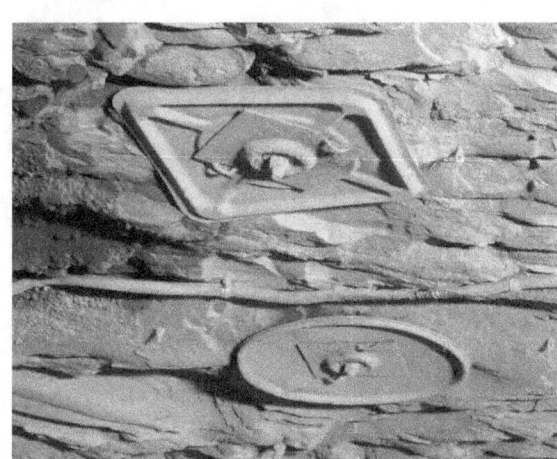

403 ft from face

Figure 55.—Roof plates at 234, 304, and 403 ft from the face after Test 5 (LLEM #505).

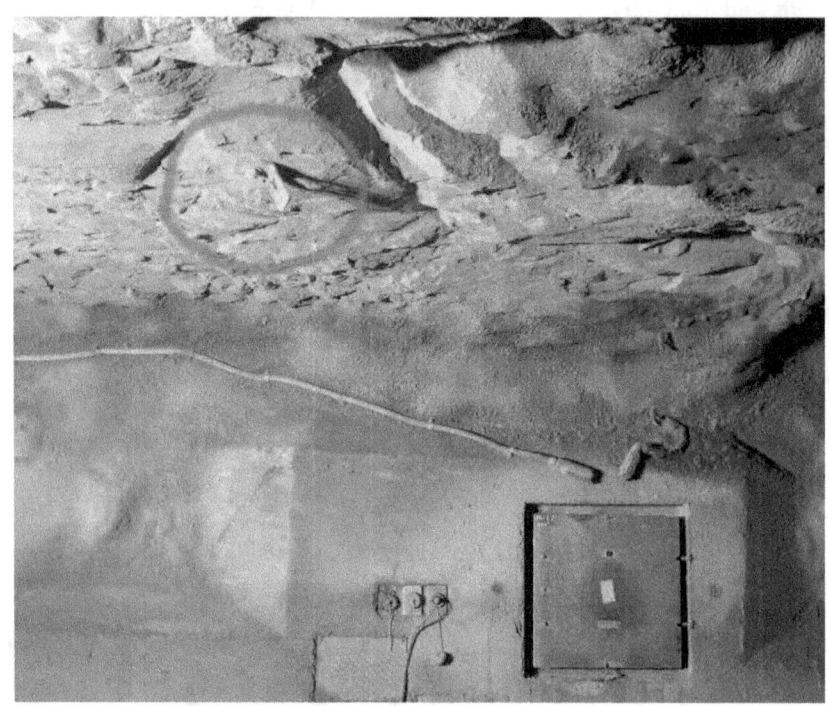

Figure 56.—Belt hanger (circled in red) near DG panel at 403 ft.

Figure 57.—Belt hanger at 403 ft after Test 5 (LLEM #505).

Test 6 (LLEM Test #506), October 19, 2006

The 2001-design Omega block seal in X-2 had survived the previous LLEM explosion tests and was left in place for the sixth and final test. A new solid-concrete-block seal was installed in X-3. This new 16-in-thick solid-concrete-block seal with a 32-in-thick center pilaster was constructed in a manner similar to the solid-concrete-block seal in X-1 for Test 1, except the X-3 seal used Type S mortar (not BlocBond) and was coated on both sides with Quikrete B-Bond sealant. Additional details on the construction of this seal can be found in Appendix B, section 6. A new Sago nominal 40-in Omega block seal was installed in C-drift about 320 ft from the face. This seal was built in a manner similar to the seals built for Test 3, except that it was built with Omega blocks from the Sago Mine. Additional details can be found in Appendix A under "Test No. 6 Protocol" and in Appendix B, section 7. Figure 58 shows a schematic of the test setup with the new Omega seal shown in blue. Wood cribs were also constructed both inby and outby the seal in C-drift, as shown in Figure 58. A hollow-block stopping was constructed outby the C-drift seal at ~384 ft. Any of the roof plates or belt hangers that had been damaged in the previous explosion were replaced for this test. Additional details of the test procedure can be found in Appendix A under "Test No. 6 Protocol."

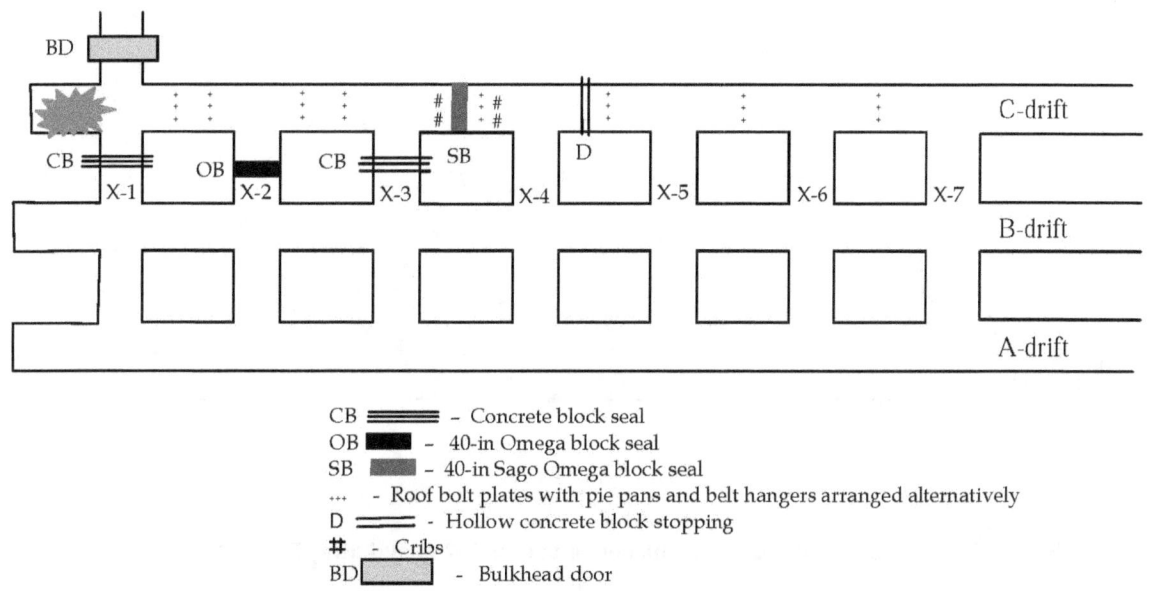

Figure 58.—Setup for Test 6 (LLEM #506).

For Test 6 (LLEM #506), a longer 71-ft gas ignition zone was used at the face of C-drift in order to generate higher pressures than those for Test 5. Test 6 was designed to exert an explosion pressure of approximately 90 psi on the Sago seal in C-drift to enable the investigators to better evaluate the explosion and seal failure that occurred at the Sago Mine. In this test, the plastic diaphragm used to confine the gas mixture was located just outby X-1. The 71-ft-long ignition zone was filled with 1,265 ft^3 of natural gas to give a mixture of ~10% CH_4 in air. Although this zone was only ~50% longer than the ignition zone for Test 5, the flammable gas volume was ~90% greater. This was due to the additional volume in X-1 between the seal and bulkhead door leading to E-drift, as shown in Figures 2 and 58. The CH_4-air zone was ignited at the face of C-drift, and the pressure pulse propagated out C-drift past the seals in X-2 and X-3 to the seal in C-drift. Because the flammable gas zone was much larger, the resulting pressures were much higher than those in Test 5.

In Test 6, the Omega block seal in X-2 and the solid-concrete-block seal in X-3 survived, but the Omega block seal in C-drift was destroyed. Figure 59 shows the pressure loading at the seal in X-2 along with the quasi-static pressure at the wall of C-drift inby the seal at 134 ft and outby the seal at 184 ft. The pressure transducer at the seal was located approximately in the middle front of the seal at 156 ft from the face of C-drift. All of the graph data were averaged over 10 ms. The smoothed pressure loading at the X-2 seal was ~51 psi. The peak pressure loading from the NI raw data at 1,500 Hz was ~68 psi at the X-2 seal, but it lasted only about 1 ms. The maximum recorded LVDT movement during the explosion was 0.06 in. Since the seal did not fail, there was essentially no change in the breakwire signal.

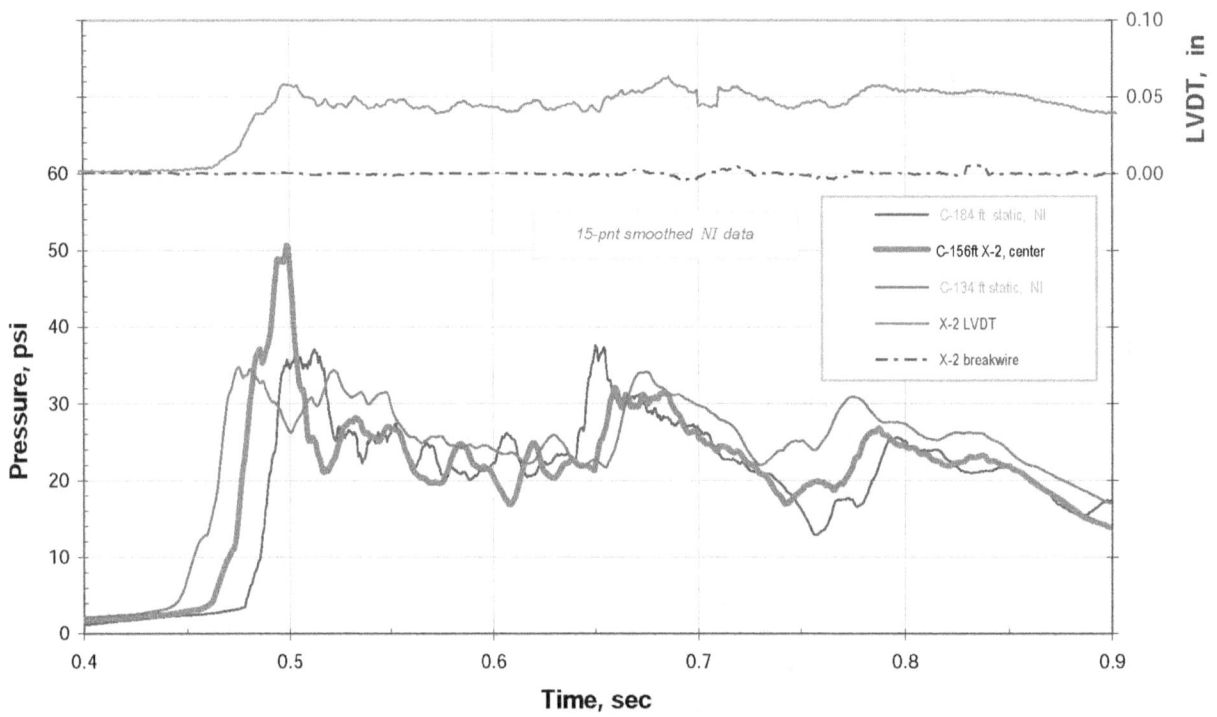

Figure 59.—Pressures and LVDT displacement at the X-2 seal during Test 6 (LLEM #506).

Figure 60 shows the pressure loadings and LVDT displacement data at the X-3 solid-concrete-block seal, along with the quasi-static pressure at the wall of C-drift inby the seal at 234 ft. The pressure transducers at the seal were located near the middle front of the seal at 256 ft from the face of C-drift. There were two pressure transducers at the seal—one mounted horizontally and one vertically, as shown in Figure 3. However, the vertical transducer did not operate properly during this test, and its data are not shown. All of the graph data were averaged over 10 ms. The seal survived both the outgoing explosion pressure loading of ~44 psi and the subsequent reflected pressure loading of ~49 psi shown in Figure 60. The peak pressure loading from the NI raw data at 1,500 Hz was ~82 psi at the X-3 seal, but it lasted less than 1 ms. The maximum recorded LVDT movement during the explosion was 0.14 in. Since the solid-concrete-block seal in X-3 did not fail, there was no change in the breakwire signals.

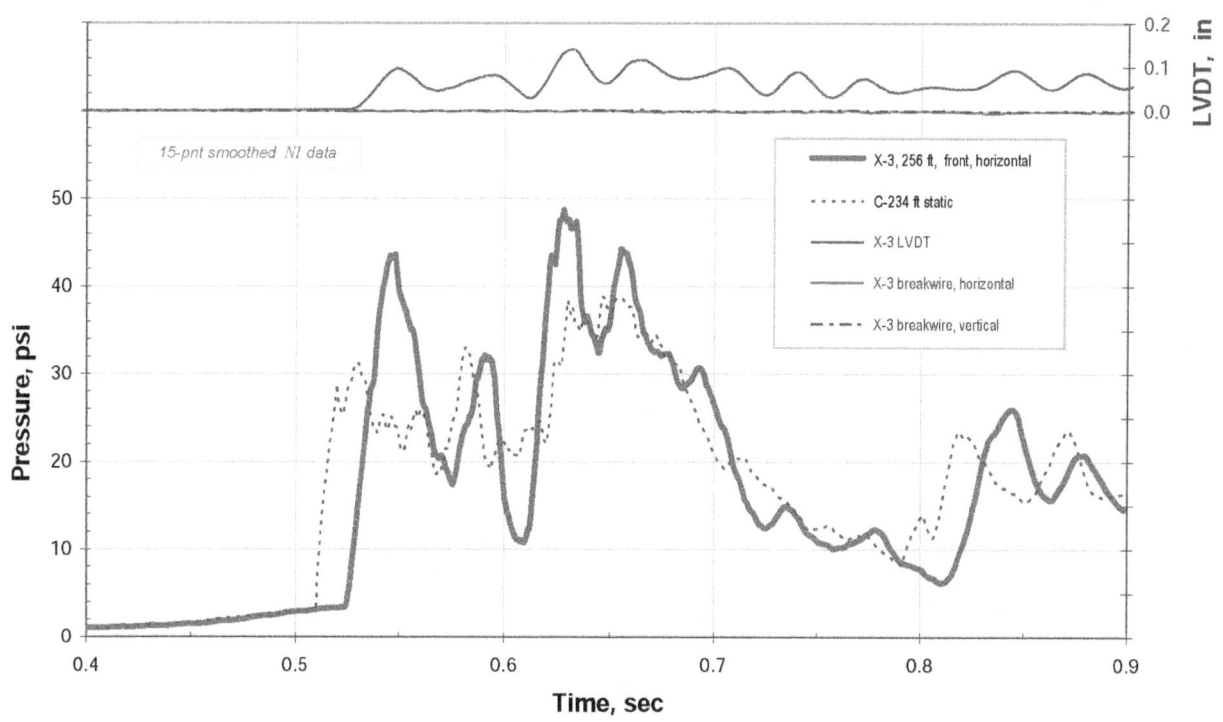

Figure 60.—Pressures and LVDT displacement at the X-3 seal during Test 6 (LLEM #506).

Figure 61 shows NI data from the pressure transducers at the C-drift seal at 320 ft from the face. Two transducers were near the middle of the seal—one mounted horizontally and one vertically, as shown in Figure 3. The third transducer was at the rib, the same as for the previous tests. For Test 6, a fourth pressure transducer was embedded into the seal near the center. This transducer was horizontal and faced the incoming pressure pulse; it therefore would read the total explosion pressure loading. The pressure traces of the three transducers near the middle of the seal were all similar, as shown in Figures 61–62. The pressure at the rib was somewhat higher. The LVDT and breakwire data in Figure 61 show that the seal broke near the time of peak pressure.

Figure 62 shows the explosion pressure loadings from the NI data acquisition system at the seal on an expanded time scale. Data for all four pressure transducers are shown. The peak smoothed (10-ms average) pressure loading for the vertical transducer was 90.1 psi. The peak smoothed pressure loading for the horizontal transducer was 91.1 psi. The peak smoothed pressure loading for the horizontal transducer that was embedded in the seal was 92.3 psi. The peak smoothed pressure loading at the rib was 99.2 psi. The LVDT data in the upper part of the figure show that the seal moved >6 in as the seal was destroyed. The two breakwire signals (from the NI raw data) both show a sharp discontinuity as the wires broke at 0.575–0.578 sec.

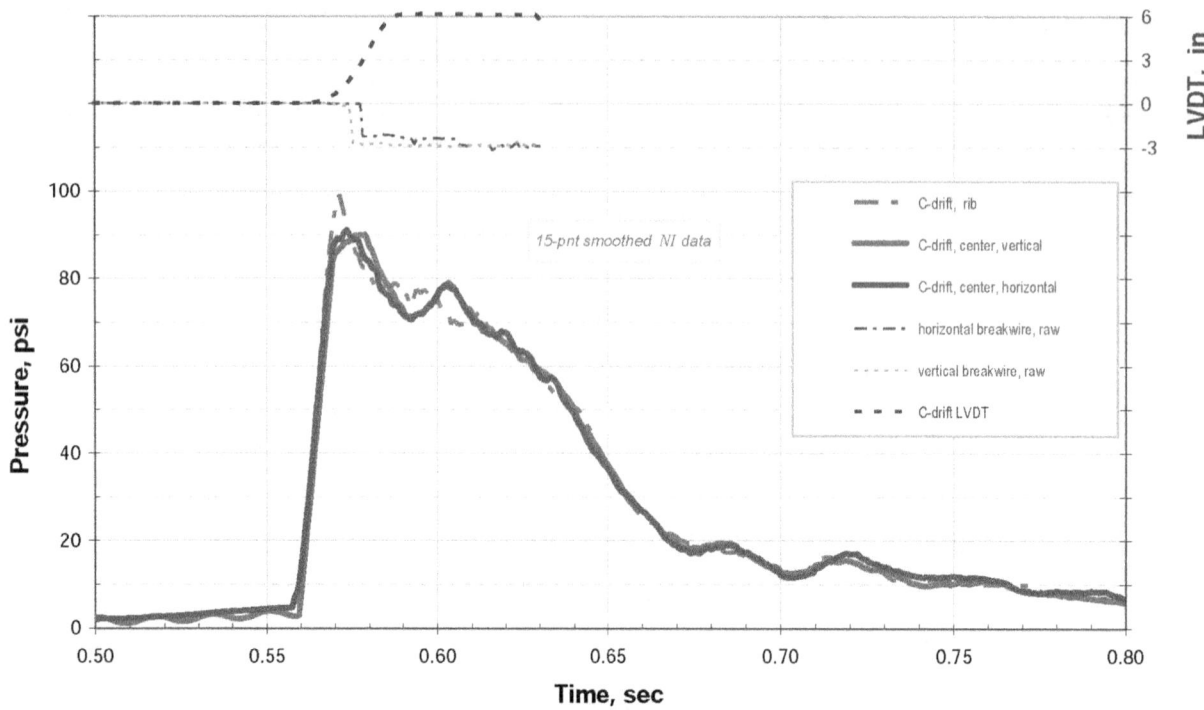

Figure 61.—Pressures and LVDT displacement at the C-drift seal during Test 6 (LLEM #506).

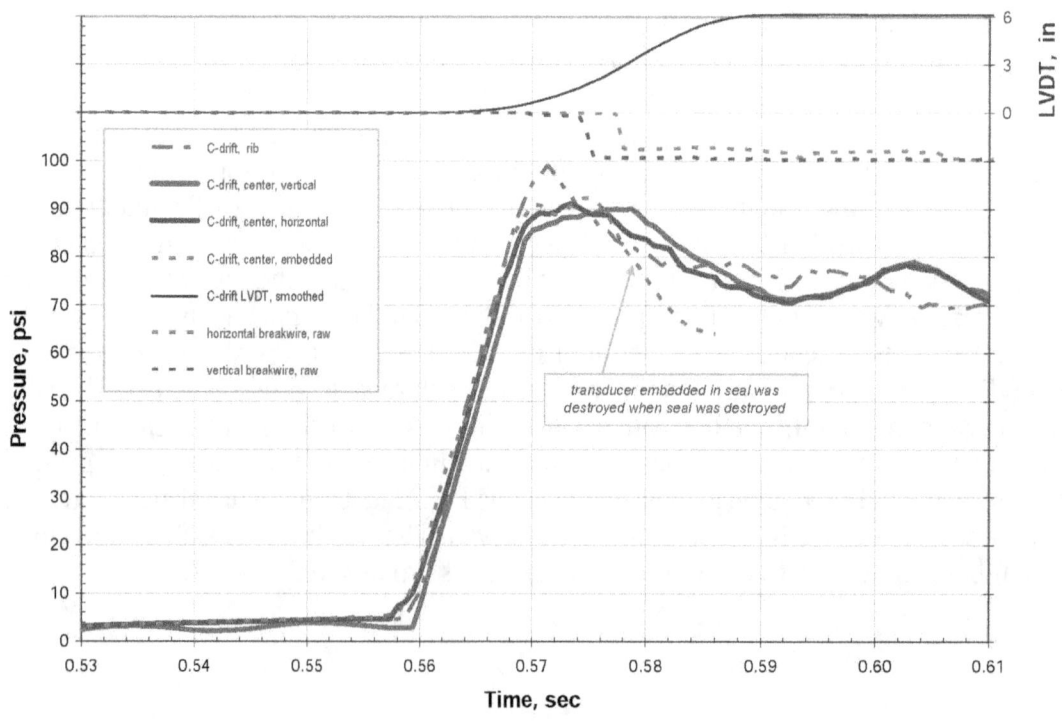

Figure 62.—Pressures and LVDT displacement on an expanded time scale at the C-drift seal during Test 6 (LLEM #506).

Figures 63–64 show the smoothed and raw data for the horizontal (on post ~1 ft in front of seal) and embedded transducers on the C-drift seal during Test 6 on an expanded time scale. The peak pressure loading from the 1,500-Hz NI raw data for the horizontal transducer was ~104 psi, and the peak from the 5,000-Hz KS raw data was ~105 psi. The peak pressure loading from the NI raw data for the embedded horizontal transducer was ~117 psi, and the peak from the KS raw data was ~120 psi. However, the horizontal transducer on the post (see Figure 3) shows higher-frequency oscillations than the embedded transducer. These oscillations may be due to mechanical vibrations. The raw data for the vertical transducer on the post show high-frequency oscillations similar to those for the horizontal transducer. The smoothed data for both transducers are almost identical.

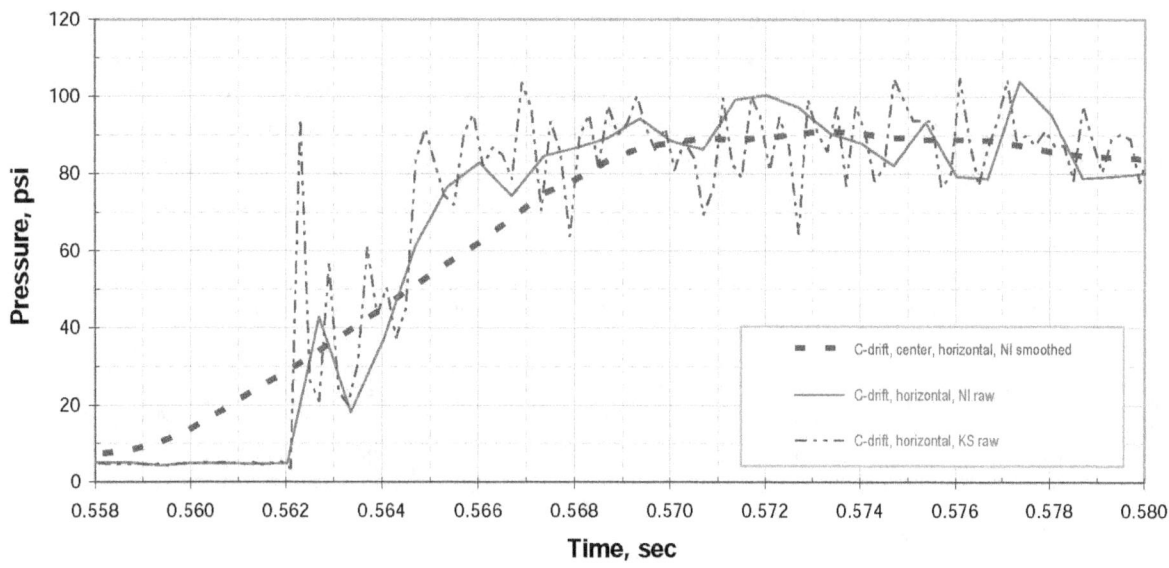

Figure 63.—Raw and smoothed pressure data for the horizontal transducer during Test 6 (LLEM #506).

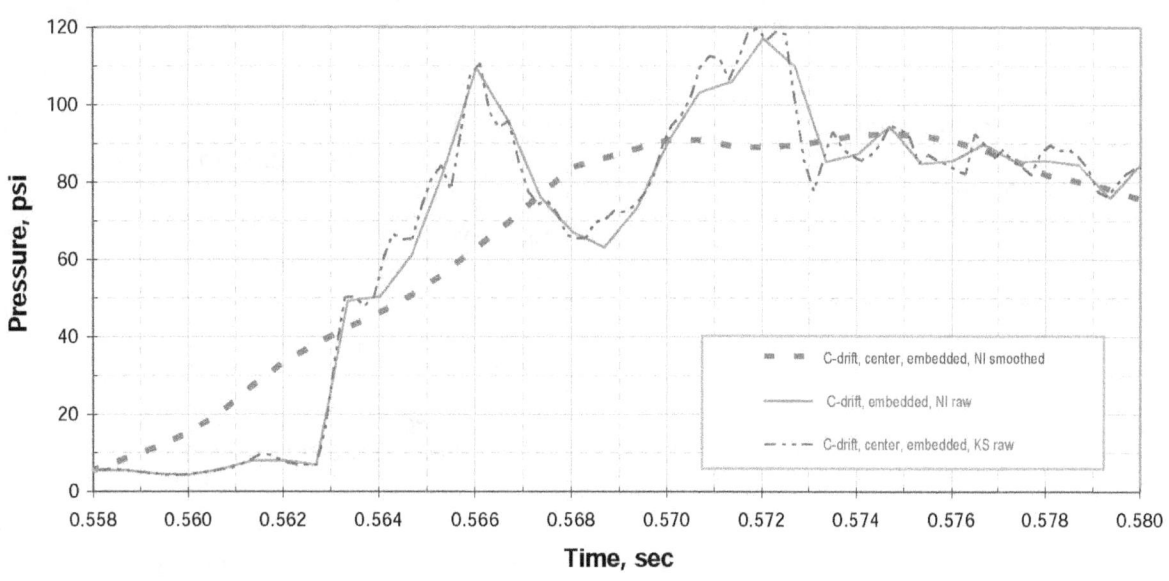

Figure 64.—Raw and smoothed pressure data for the embedded transducer during Test 6 (LLEM #506).

Pressure, LVDT displacement, and breakwire summary data for the seals in X-2, X-3, and C-drift during Test 6 are shown in Table 11. The peak smoothed pressure loadings at the seals are listed for the NI and KS data acquisition systems. The data were averaged over 10 ms for the listed pressure values. The breakwire time is from the NI raw data. The peak smoothed pressure loading at the middle of the X-2 Omega block seal was 50.7 psi and 49.2 psi from the NI and KS data acquisition systems, respectively, and the seal survived the explosion. The peak smoothed pressure loading from the horizontal transducer at the middle of the solid-concrete-block X-3 seal was 48.8 psi and 48.4 psi from the NI and KS systems, respectively, and the seal survived.

The peak smoothed pressure loading from the horizontal and vertical transducers on the post at the middle of the C-drift seal was 89–91 psi from the NI and KS data acquisition systems. The transducer embedded in the seal saw a slightly higher pressure loading of ~92.5 psi from both data acquisition systems. The pressure loading at the rib was ~99 psi from both data acquisition systems. The breakwire raw data show that the vertical wire broke at 0.575 sec and the horizontal wire broke at 0.578 sec as the seal was destroyed. The pressure loading at the stopping in C-drift (point D in Figure 58), as measured from the 384-ft transducer located in front of the stopping, was ~7.5 psi, and the stopping was totally destroyed. The breakwire raw data show that the stopping was destroyed at 0.79 sec.

Table 11.—Pressures and LVDT displacement at the seals during Test 6 (LLEM #506)

Seal or stopping		Seal pressures		LVDT deflection,	Break time,	Seal or stopping		Seal pressures		LVDT deflection,	Break time,
		psi (NI)	psi (KS)	in	sec			psi (NI)	psi (KS)	in	sec
X-2 BC seal 156 ft	H	50.7	49.2	0.06							
		Seal survived									
X-3 BC seal 256 ft	H	48.8	48.4	0.14	—						
	V	—	—		—						
						Wood cribs	A				0.561
		Seal survived					B				0.562
								Cribs destroyed			
						C-drift seal 320 ft	Rib	99.2	99.3	>6	
							H,e	92.3	92.6		
							H	91.1	89.0		0.578
							V	90.1	89.2		0.575
								Seal destroyed, Omega block debris traveled ~917 ft, piece of BlocBond traveled ~1,168 ft			
						C-drift stopping 384 ft	H	7.5	7.6		0.790
								Stopping destroyed, debris traveled ~748 ft			

H Horizontal.   V Vertical.   e Embedded.

Table 12 lists quasi-static pressure and flame sensor data at the various DG panels on the walls of B- and C-drifts during LLEM #506. The positions of the Omega block seals in X-2 and C-drift are depicted by blue shading. On the left of the table are the peak smoothed B-drift quasi-static wall pressures; toward the right are the peak smoothed quasi-static wall pressures in C-drift. The C-drift pressures were relatively constant from 84 ft out to 234 ft. The pressure increased significantly out to ~320 ft as the pressure pulse was confined by the C-drift seal. After the seal broke, the pressures beyond the seal were much lower: 7.5 psi at the stopping at point D (384 ft from closed end of C-drift as shown in Figure 58), ~4.6 psi at 403 ft, and ~3 psi at 501 ft, etc. The last two columns of Table 12 list the flame signal and arrival time at each of the DG panels. For this test, the flame went past the 234-ft panel but did not reach the 304-ft panel. Therefore, the interpolated flame travel distance was about 240 ft. Based on the initial gas zone length of 71 ft, the expansion ratio would be about 3.4 (240 ft/71 ft).

Table 12.—Wall pressures and flame travel during Test 6 (LLEM #506)

B-drift quasi-static pressures				C-drift quasi-static pressures			Flame signal	
Distance, ft	psi (NI)	psi (KS)		Distance, ft	psi (NI)	psi (KS)	Volts	sec (NI)
				3	42.4	42.5		
10	5.6	—	1	13	—	—	>5	0.227
			X-1					
108	4.4	—	2	84	36.6	36.7	>5	0.415
158	~4	—	3	134	34.7	35.0	>5	0.454
			X-2					
211	—	—	4	184	37.6	37.6	>5	0.510
257	2.4	—	5	234	38.9	38.9	>5	0.572
			X-3					
329	2.7	—	6	304	89.5	88.1	~0	
			X-4					
427	3.2	—	7	403	4.6	4.7	~0	
			X-5					
526	3.3	—	8	501	3.2	2.7	~0	
			X-6					
626	3.1	—	9	598	3.0	3.1	~0	
			X-7					
782	3.0	—	10	757	3.0	3.0	~0	
			11	1,506	2.3	2.2		

The Omega block seal in X-2 and the solid-concrete-block seal in X-3 survived the explosion during Test 6 (LLEM #506). The air leakage data are in Table C-10 in Appendix C. Both seals passed the leakage test. The seal in C-drift was destroyed during the test and therefore not measured for air leakage.

The debris from the C-drift seal was thrown a long distance down the drift. Figure 65 shows the postexplosion view looking outby near the original seal position in C-drift. Near the center is the pressure transducer post. The original seal location was just behind this post. The vertical transducer is still attached to the back of the post, but the horizontal transducer was broken from the post during the explosion. The BDP at 306 ft is in the left center of the photo. Only a few wood blocks and small pieces of seal debris are seen on the floor outby the original seal position.

Figure 65.—Postexplosion view in C-drift looking outby toward original seal location at 320 ft after Test 6 (LLEM #506).

Figure 66 shows the view looking outby from X-4 at 355 ft. Only a few pieces of wood and seal debris are seen in the photo. Figure 67 shows the view looking outby from the DG panel at 403 ft toward X-5 at 451 ft. At this distance, there are still only a few pieces of wood and seal debris. Figure 68 shows the view looking outby from ~500 ft toward X-6 at 547 ft, while Figure 69 shows the view looking outby from ~600 ft. Both figures show only small amounts of debris from the wood cribs, C-drift seal, and C-drift stopping.

Figure 66.—Postexplosion view in C-drift looking outby from X-4 at 355 ft.

Figure 67.—Debris in C-drift after Test 6 (LLEM #506) looking outby from 403 ft.

Figure 68.—Debris in C-drift after Test 6 (LLEM #506) looking outby from ~500 ft.

Figure 69.—Debris in C-drift after Test 6 (LLEM #506) looking outby from ~600 ft.

Figure 70 shows the debris looking outby in C-drift from the DG panel at 757 ft from the face or ~435 ft from the original seal location. The DG panel is on the left edge of the photo. There are numerous wood blocks and pieces of seal debris in the photo. Figure 71 shows the view looking outby from ~850 ft. Pieces of wood and seal debris are more numerous at this distance.

Figure 70.—Debris in C-drift after Test 6 (LLEM #506) looking outby from 757 ft.

Figure 71.—Debris in C-drift after Test 6 (LLEM #506) looking outby from ~850 ft.

Figure 72 shows a photo of the postexplosion debris looking outby in C-drift from ~950 ft from the face or ~630 ft from the original seal location. At this distance, large amounts of wood and seal debris were found.

Figure 72.—Debris after Test 6 (LLEM #506) looking outby from ~950 ft.

Figure 73 shows the view looking outby in C-drift from ~1,050 ft after Test 6 (LLEM #506). There were large amounts of wood and seal debris at this location also. Figure 74 shows the view looking outby from ~1,150 ft from the face or ~825 ft from the original seal location. There was less debris at this location.

Figure 75 shows the view looking outby in C-drift from ~1,240 ft after Test 6. There were a few wood blocks and some pieces of seal debris at this location. Figure 76 shows the view looking outby from ~1,440 ft from the face or ~1,120 ft from the original seal location. The wood block was debris from the wood cribs that were outby the C-drift seal at the start of the test. The piece of seal debris (to the left and beyond the wood block) was a piece of BlocBond mortar. These were the debris pieces that traveled the farthest in C-drift during explosion Test 6 (LLEM #506).

Figure 73.—Debris after Test 6 (LLEM #506) looking outby from ~1,050 ft.

Figure 74.—Debris after Test 6 (LLEM #506) looking outby from ~1,150 ft.

Figure 75.—Debris in C-drift after Test 6 (LLEM #506) looking outby from ~1,240 ft.

Figure 76.—Debris in C-drift after Test 6 (LLEM #506) looking outby from ~1,440 ft.

During Test 6 (LLEM #506), pieces of Omega blocks traveled as far as ~917 ft from the original C-drift seal location, as noted in Table 11. A piece of BlocBond mortar traveled ~1,168 ft from the original seal location (see Figure 76). Debris from the C-drift stopping traveled ~748 ft from the original stopping location. Debris from the wood cribs that were originally inby the C-drift seal traveled ~606 ft. Debris from the wood cribs that were originally outby the seal traveled ~1,139 ft (see Figure 76). These distances were measured from the postexplosion surveys.

For Test 6, the 1,560-lb battery charger was placed at 688 ft from the face of C-drift or 365 ft from the outby face of the C-drift seal. The charger moved 356 ft during Test 6 to a location ~1,044 ft from the face (see Figure 77). The total explosion pressure during Test 6 at the 604-ft BDP was 3.0 psi and the dynamic pressure was 0.3 psi. The peak smoothed quasi-static wall pressure at 757 ft was 3.0 psi. The peak smoothed pressure loading at the charger would have been ~3 psi. However, for Test 6, the pressure pulse took longer than 7 ms to reach its maximum value. Therefore, for this test, the maximum differential pressure loading from the inby to outby end of the charger was only ~1 psi. The cross-sectional area of 900 in^2 and differential explosion pressure loading of ~1 psi would result in a total force of ~900 lb for a few milliseconds. The 0.3-psi dynamic pressure (~270-lb force) would have continued to act on the charger for a longer period of time. In addition to the air pressure, the battery charger was also hit by debris from the seal, stopping, and wood cribs (see Figure 77).

Figure 77.—Final position of battery charger after Test 6 (LLEM #506).

Some of the roof plates were damaged during the Test 6 explosion, as shown in Figures 78–80. On the left side of the figures, the plates are shown from inby of the plates, looking outby. On the right side of the figure, the plates are viewed looking across the entry from the DG-panel side. There was only a slight bending of the round plate at 84 ft. The round plate at 134 ft from the face was severely bent during the explosion, but the square plate had no obvious damage.

Viewed from inby	Viewed across entry from DG panel

84 ft from face

134 ft from face

Figure 78.—Roof plates at 84 and 134 ft from the face after Test 6 (LLEM #506).

At 184 and 234 ft, both the round and square roof plates were severely bent by the explosion, as shown in Figure 79. Two square roof plates instead of just one were inadvertently installed at the 184-ft location. The round plate at 234 ft was rotated approximately 180° during the explosion. The quasi-static wall pressures as measured by the sensors in the DG panels were ~38 psi at the 184-ft panel and ~39 psi at the 234-ft panel. In Figure 80, both the round and square plates at 304 ft were severely bent by the explosion; the quasi-static wall pressure at this location was ~89.5 psi. At 403 ft, both the round and square plates were bent at the edges, possibly from being hit by flying debris. The round plate at 403 ft rotated about one-quarter turn during the explosion. The quasi-static wall pressure at this location was ~4.6 psi. There was little or no obvious damage to the other round or square plates throughout C-drift.

Figure 79.—Roof plates at 184 and 234 ft from the face after Test 6 (LLEM #506).

Viewed from inby	Viewed across entry from DG panel

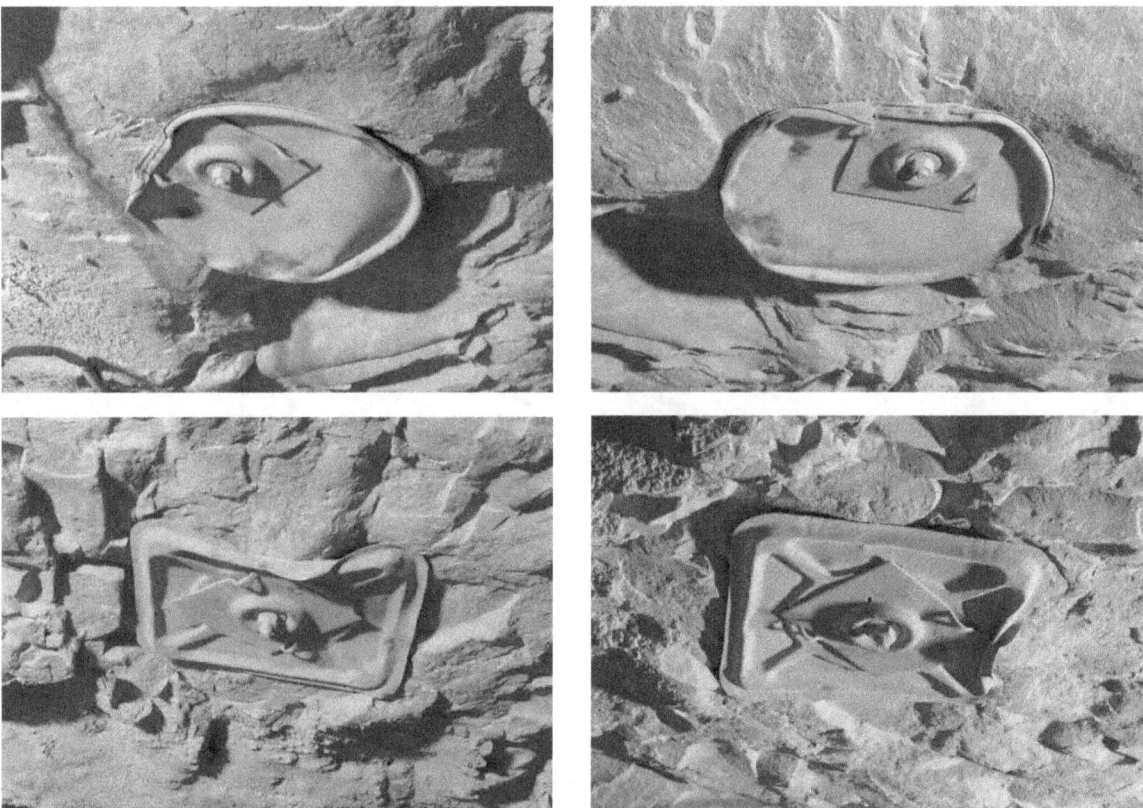

304 ft from face

403 ft from face

Figure 80.—Roof plates at 304 and 403 ft from the face after Test 6 (LLEM #506).

The belt hanger at 403 ft was significantly bent, as shown in Figures 81–82. This was similar to the damage to this belt hanger during Test 5. These views are looking across the entry toward the DG panel. Therefore, outby is to the right in these photos. This belt hanger was probably damaged by flying debris because the inby belt hangers were not damaged even though they were exposed to greater explosion pressures. The belt hanger seemed to have a small amount of residue from the impact of an Omega block. For this test, the carbon monoxide sensor box was hung from the roof at ~800 ft from the face or ~475 ft from the seal. There was no apparent damage to the sensor box during the explosion.

Figure 81.—Belt hanger (circled in red) near DG panel at 403 ft.

Figure 82.—Belt hanger at 403 ft after explosion Test 6 (LLEM #506).

SUMMARY AND CONCLUSIONS

Several seal designs using Omega 384 blocks were constructed at the LLEM during 2006 and exposed to various explosion pressures. All of the seals were constructed of Omega low-density blocks with nominal dimensions of 8 in by 16 in by 24 in. The blocks were alternated to stagger the joints. In the 2001 design, properly mixed BlocBond mortar was applied to all of the block-to-block and block-to-strata interfaces, including the floor. There were some differences between the 2001 design and the hybrid and Sago designs. The main differences between the hybrid design and the 2001 design were that the hybrid design was installed on a 0.25-in-thick layer of dry BlocBond and the entire first course of blocks was put into position before applying any mortar to the blocks. For all subsequent courses with the hybrid design, the mortar was applied by gloved hand to the block joints before placement of each block. The main differences between the Sago design and the 2001 design were that the Sago design was installed on a 1.5-in-thick layer of dry BlocBond and the mortar was forced into the vertical joints after the blocks were positioned for all courses of blocks. Comprehensive details of the three seal construction procedures can be found in Appendix B.

A summary of the results of the explosions against the three seal designs is presented in Table 13. The first two columns of the table list the type of seal design and location in the LLEM. The next two columns list the seal height and width. All of the seals were nominally 40 in thick. When the coating thickness on the faces of the seal and the mortar thickness are included, the total seal thickness was about 41 in. The next column lists the smoothed explosion pressure loading at which a particular seal survived. The final column lists the smoothed explosion pressure loading at which a particular seal was destroyed. This value is the peak pressure measured during a particular explosion at the middle front of the seal. If a particular design of seal was destroyed during more than one explosion, the lower pressure loading is listed. For example, a Sago seal in C-drift was destroyed at 57 psi during Test 5 and at 93 psi during Test 6, so only the lower pressure loading of 57 psi is listed in Table 13. The strength of a particular seal would be somewhere between the values in columns 5 and 6 of the table. For example, the 81-in-high hybrid seal survived a pressure loading of 25 psi and was destroyed during a later pressure loading of 39 psi. Therefore, its strength is greater than 25 psi but less than 39 psi, when comparing the 10-ms peak pressure averages.

Table 13.—Summary of explosion pressures on various seals

Seal design	Location	Height, in	Width, in	Pressure loading at which seal survived, psi	Pressure loading at which seal was destroyed, psi
2001	X-2	80	226	51	NA
2001	C-drift	88	224	NA	51
Hybrid	X-3	81	226	25	39
Sago	X-3	80	226	18	35
Sago	C-drift	88	224	21	57

NA   No data were available for this scenario.

The 2001-design Omega block seal (see Appendix B, section 1 for construction details) located in X-2 survived all six LLEM explosions, with peak pressure loadings of 13, 15, 22, 23, and 51 psi. Note that all of the pressure loading values were smoothed data that were averaged over 10 ms. The pressure data here are all from transducers near the geometric center in front of the seals. The 2001-design Omega block seal (Appendix B, section 3) in C-drift was destroyed during an explosion (Test 2) with a pressure loading of 51 psi. The differences in heights between these two seals and the orientation of each seal to the explosion were contributing factors to the fact that the X-2 seal survived Test 6 at 51 psi and the C-drift seal was destroyed during Test 2 at 51 psi. The C-drift seal was ~88 in high and the X-2 seal was ~80 in high. The higher seal would be weaker for the same seal thickness [Anderson 1984]. In addition, the X-2 seal was subjected to a nonuniform, sweeping pressure loading while the C-drift seal was subjected to a more uniform, head-on explosion pressure loading. The hybrid Omega block seal (Appendix B, section 2) in X-3 survived Test 1 at a pressure loading of 25 psi and failed during Test 2 at a pressure loading of 39 psi. Based on these LLEM tests, the hybrid seal design is weaker than the 2001 seal design.

The Sago Omega block seals were constructed in X-3 and C-drift before Test 3, as described in Appendix B, sections 4 and 5. The X-3 seal survived Tests 3 and 4 at pressure loadings of 16 psi and 18 psi, respectively. It was destroyed during Test 5 at a pressure loading of 35 psi. The C-drift seal survived head-on explosions (Tests 3 and 4), which resulted in pressure loadings of 17 psi and 21 psi, respectively. It was destroyed during Test 5 at a pressure loading of 57 psi. The X-3 and C-drift Sago Omega block seals both survived Test 4, which generated a pressure loading of 18 psi at the X-3 seal and a pressure loading of 21 psi at the C-drift seal. The X-3 and C-drift seals both were destroyed during Test 5 at higher pressure loadings of 35 psi and 57 psi, respectively. Another Sago Omega block seal was constructed across C-drift for Test 6; it was destroyed by a pressure loading of 93 psi, as expected. Based on these LLEM tests, the Sago seal design is weaker than the 2001 seal design.

During these LLEM explosion tests, the distance of seal debris travel was also measured. The C-drift seal was exposed to a pressure loading of 51 psi in Test 2, and the seal debris was thrown over 800 ft. In Test 2, there was no significant obstacle beyond the C-drift seal that would restrict the debris travel. In Tests 5 and 6, there were two wood cribs and a hollow-concrete-block stopping beyond the C-drift seal. Even though the cribs and stopping were destroyed in both tests, they would absorb blast energy and therefore limit the debris travel distance. In Test 5, the C-drift seal was exposed to a pressure loading of 57 psi, and the seal debris was thrown over 500 ft. In Test 6, the C-drift seal was exposed to a pressure loading of 93 psi, and the Omega block debris was thrown over 900 ft. During these LLEM tests, the explosion pressure effects on other structures and objects were also documented, as described earlier in this report.

The purpose of these LLEM explosion tests conducted in 2006 was to assist MSHA and WVOMHS&T in determining the explosion pressure loadings at which various 40-in-thick Omega block seal designs would fail relative to LLEM conditions and in studying the explosion effects on various mine items, including the debris fields resulting from the destroyed seals. The information in this report was used as supporting data in the WVOMHS&T and MSHA analyses of the Sago Mine explosion and their subsequent investigative reports [WVOMHS&T 2006; Gates et al. 2007].

ACKNOWLEDGMENTS

The authors acknowledge the following individuals for the invaluable technical support they provided during the development of the test protocols and in the postexplosion observations and analyses: Richard A. Gates, MSHA District 11 Manager and the lead Sago Mine accident investigator; Clete R. Stephan, Principal Mining Engineer, MSHA Ventilation Division; Monte Hieb, P.E., Chief Engineer, WVOMHS&T; J. Brian Mills, District 1 Inspector, WVOMHS&T; and Jürgen F. Brune, Ph.D., former Chief, Disaster Prevention and Response Branch, NIOSH-PRL. We also acknowledge John Cruse, Technical Analyst, WVOMSH&T, for his contributions to the postexplosion observations and documentation. We recognize Benjamin Singleton, Michael Keener, Stacy Brown, and Damon Wilkewitz of Allegheny Surveys, Inc., Birch River, WV, for their timely and accurate surveys of the debris and objects after each LLEM explosion test and the maps they provided of the seal locations and debris fields. In addition, we acknowledge Charles Lash, Technical Services Representative, and Michael Shook, Mining Engineer, of Burrell Mining Products, Inc., New Kensington, PA, for providing the Omega 384 blocks and for their efforts in the construction of the 2001-design Omega block seal in crosscut 2.

We thank the following NIOSH-PRL personnel without whose contributions the LLEM seal evaluation program could not have been accomplished: William A. Slivensky, Frank A. Karnack, and Donald D. Sellers, Physical Science Technicians at Lake Lynn Laboratory, for their participation in the installation of sensors for each seal and stopping design, construction of the methane ignition zones and installation of the gas sampling lines, systems checks and shotfire during each test, postexplosion observations, and documentation of other measurements; Kenneth W. Jackson, Electronics Technician at Lake Lynn, for sensor calibrations, operation of the data acquisition systems, and initial data analyses; and Cynthia A. Hollerich, Physical Science Technician at PRL, for her detailed documentation of the seal installations and postexplosion observations.

We acknowledge the following mechanical technicians with Ki Corp. (a NIOSH contractor) for their extensive efforts in the installation of the Omega block seal designs, solid-concrete-block seal, and hollow-core block stoppings; the thorough and timely cleanup of debris after each test; and the washdown of the entry: James D. Addis (now a Physical Science Technician with PRL), Timothy Glad, James Rabon, and Bernard Lambie.

All photographs in this report were taken by Cynthia A. Hollerich, Kenneth L. Cashdollar, and Eric S. Weiss of NIOSH-PRL.

REFERENCES

Anderson C [1984]. Arching action in transverse laterally loaded masonry wall panels. Struct Eng *62B*(1):22.

CFR. Code of federal regulations. Washington DC: U.S. Government Printing Office, Office of the Federal Register.

Gates RA, Phillips RL, Urosek JE, Stephan CR, Stoltz RT, Swentosky DJ, Harris GW, O'Donnell JR Jr., Dresch RA [2007]. Report of investigation: fatal underground coal mine explosion, January 2, 2006. Sago mine, Wolf Run Mining Company, Tallmansville, Upshur County, West Virginia, ID No. 46-08791. Arlington, VA: U.S. Department of Labor, Mine Safety and Health Administration.

Greninger, NB, Weiss ES, Luzik SJ, Stephan CR [1991]. Evaluation of solid-block and cementitious foam seals. Pittsburgh, PA: U.S. Department of the Interior, Bureau of Mines, RI 9382. NTIS No. PB 92-152115.

Mattes RH, Bacho A, Wade LV [1983]. Lake Lynn Laboratory: construction, physical description, and capability. Pittsburgh, PA: U.S. Department of the Interior, Bureau of Mines, IC 8911. NTIS No. PB 83-197103.

Nagy J [1981]. The explosion hazard in mining. Pittsburgh, PA: U.S. Department of Labor, Mine Safety and Health Administration, IR 1119.

Nagy J, Mitchell DW [1963]. Experimental coal-dust and gas explosions. Pittsburgh, PA: U.S. Department of the Interior, Bureau of Mines, RI 6344.

Sapko MJ, Weiss ES, Trackemas J, Stephan CR [2004]. Designs for rapid in situ sealing. In: Yernberg WR, ed. Transactions of Society for Mining, Metallurgy, and Explorations, Inc. Vol. 316. Littleton, CO: Society for Mining, Metallurgy, and Exploration, Inc., pp. 85–92.

Stephan CR [1990a]. Construction of seals in underground coal mines. Pittsburgh, PA: U.S. Department of Labor, Mine Safety and Health Administration, Industrial Safety Division (ISD) report No. 06-213-90, August 1, 1990.

Stephan CR [1990b]. Omega 384 block as a seal construction material. Pittsburgh, PA: U.S. Department of Labor, Mine Safety and Health Administration, Industrial Safety Division (ISD) report No. 10-318-90, November 14, 1990.

Triebsch G, Sapko MJ [1990]. Lake Lynn Laboratory: a state-of-the-art mining research laboratory. In: Proceedings of the International Symposium on Unique Underground Structures. Vol. 2. Golden, CO: Colorado School of Mines, pp. 75-1 to 75-21.

Weiss ES, Greninger NB, Perry JW, Stephan CR [1993a]. Strength and leakage evaluations for coal mine seals. In: Proceedings of the 25th International Conference of Safety in Mines Research Institutes (Pretoria, South Africa, September 13–17, 1993), Conference Papers for Day One, pp. 149-161.

Weiss ES, Greninger NB, Slivensky WA, Stephan CR [1993b]. Evaluation of alternative seal designs for coal mines. In: Proceedings of the Sixth U.S. Mine Ventilation Symposium (Salt Lake City, UT, June 21–23, 1993). Chapter 97. Littleton, CO: Society for Mining, Metallurgy, and Exploration, Inc., pp. 635–640.

Weiss ES, Greninger NB, Stephan CR, Lipscomb JR [1993c]. Strength characteristics and air-leakage determinations for alternative mine seal designs. Pittsburgh, PA: U.S. Department of the Interior, Bureau of Mines, RI 9477. NTIS No. PB94111275.

Weiss ES, Slivensky WA, Schultz MJ, Stephan CR, Jackson KW [1996]. Evaluation of polymer construction material and water trap designs for underground coal mine seals. Pittsburgh, PA: U.S. Department of Energy, RI 9634. NTIS No. PB96-123392.

Weiss ES, Slivensky WA, Schultz MJ, Stephan CR [1997]. Evaluation of water trap designs and alternative mine seal construction materials. In: Dhar BB, Bhowmick BC, eds. Proceedings of the 27th International Conference of Safety in Mines Research Institutes (New Delhi, India, February 20–22, 1997). Vol. II. New Delhi, India: Oxford & IBH Publishing Co. Pvt. Ltd., pp. 973–981.

Weiss ES, Cashdollar KL, Mutton IVS, Kohli DR, Slivensky WA [1999]. Evaluation of reinforced cementitious seals. Pittsburgh, PA: U.S. Department of Health and Human Services, Centers for Disease Control and Prevention, National Institute for Occupational Safety and Health, DHHS (NIOSH) Publication No. 99–136, RI 9647.

WVOMHS&T (West Virginia Office of Miners' Health, Safety, and Training) [2006]. Report of investigation into the Sago mine explosion which occurred January 2, 2006, Upshur Co., West Virginia. Charleston, WV: West Virginia Office of Miners' Health, Safety, and Training.

Zipf RK Jr., Sapko MJ, Brune JF [2007]. Explosion pressure design criteria for new seals in U.S. coal mines. Pittsburgh, PA: U.S. Department of Health and Human Services, Centers for Disease Control and Prevention, National Institute for Occupational Safety and Health, DHHS (NIOSH) Publication No. 2007–144, IC 9500.

APPENDIX A.—MSHA–WVOMHS&T–NIOSH PROTOCOLS FOR THE LLEM EXPLOSION TESTS

This appendix contains the protocols from the LLEM explosion tests conducted in 2006 as developed by MSHA, WVOMHS&T, and NIOSH. These protocols are reproduced on the following pages in their original format.

NIOSH - MSHA- WVOMHS&T Seal Testing
Test No. 1 Protocol -- LLEM test #501

MSHA and WVOMHS&T are planning a series of tests at NIOSH's Lake Lynn Laboratory, Experimental Mine, to evaluate 40-inch thick Omega Block Seals. These tests are being performed as a direct result of the fatal explosion that occurred at the Sago Mine on January 2, 2006, where ten 40-inch thick Omega Block Seals were destroyed. Individual tests, within this series, will be designed to address the results of preceding tests, requiring a separate protocol for each test. This series of tests will assist investigators in determining why 40-inch thick Omega Block Seals failed at the Sago Mine.

Based on preliminary information, the following may have occurred at the Sago Mine:
1) Explosion pressures may have exceeded 20 psi;
2) The seals may not have been properly constructed;
3) Construction materials may have been substandard;
4) Cribs inby the seals may have contributed to the failure of the seals; and
5) A combination of any or all of the four items listed above.

Test No. 1:
A 40-inch thick Omega Block Seal was constructed as tested and approved in 2001. A second 40-inch Omega Block Seal was constructed with certain modifications, as identified below. This hybrid 40-inch thick Omega Block Seal was not intended to be a duplicate of the seals at the Sago Mine. However, it includes some similarities. Additional 40-inch thick Omega Block Seals will be constructed and tested in the future based on additional information including that obtained from this initial explosion test and from other engineering tests and evaluations.

Purpose:
This first explosion test will attempt to evaluate the results of a 20 psi static pressure pulse from an explosion on a 40-inch thick Omega Block Seal, which was constructed in the same manner as the seals tested in 2001, and on the hybrid 40-inch thick Omega Block Seal. This test will be conducted 22 days after completion of the seals.

This test is not intended to duplicate the explosion and seal failure that occurred at the Sago Mine, and the results should not be interpreted as a replication of those events. This test is intended to provide a basis for future seal tests that may offer insight about the Sago Mine seal failures.

Method and Protocol:
The 40-inch thick Omega Block Seal was constructed as tested and approved in 2001. This seal was built by Burrell Mining Products personnel. NIOSH personnel documented the construction techniques.

The hybrid 40-inch thick Omega Block Seal was constructed by NIOSH contractors under the direction of MSHA and WVOMHS&T personnel using construction techniques as the approved seal, with three exceptions. These exceptions include; applying unmixed mortar

NIOSH - MSHA- WVOMHS&T Seal Testing
Test No. 1 Protocol -- LLEM test #501

on the mine floor, not applying mortar directly to the vertical joints of the first course of blocks, and modifying the installation of wood planks and wedges between the last course of the Omega Blocks and the mine roof.

After a 22-day curing period, the seals will be subjected to an approximately 20 psi static pressure pulse from an explosion.

The openings in the two final seals at the Sago Mine were completed on December 11, 2005, which was 22 days before the explosion occurred. Test No. 1 will be conducted 22 days after completion of the seals to replicate the shortest curing period on any portion of the 40-inch thick Omega Block Seals at the Sago Mine.

NIOSH - MSHA - WVOMHS&T Seal Testing
Test No. 1 Protocol -- LLEM test #501

Test No. 1

1) Install a Solid Concrete Block Seal in the 1st Crosscut.
2) Have Burrell Mining Products install a 40-inch thick Omega Block Seal in the 2nd Crosscut between the Nos. 2 and 3 Entries according to the approved 2001 methods. (See Attachment 1)
3) Install a hybrid 40-inch thick Omega Block Seal in the 3rd Crosscut between the Nos. 2 and 3 Entries. (See Attachment 2)
4) Install 2 cribs in the No. 3 Entry outby the No. 3 Crosscut.
5) Install belt hangers along the roof in the No. 3 Entry and roof bolt bearing plates with pie pans along the roof in the No. 3 Entry. Locate hangers and plates alternatively along the entry.
6) Record all pre-explosion parameters, including volume and percentage of methane.
7) Test with an approximately 20 psi static pressure pulse from an explosion 22 days after completion.
8) Determine overpressures throughout the test area.
9) Record and map all post explosion results, including overpressures and flame length.

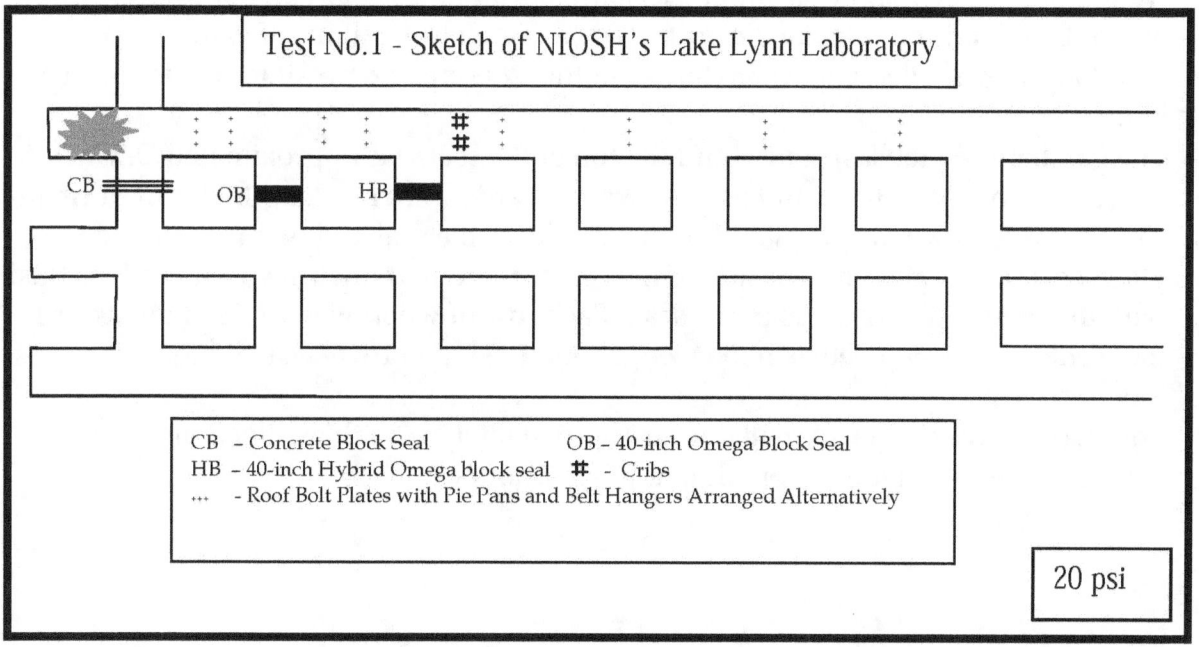

Test No.1 - Sketch of NIOSH's Lake Lynn Laboratory

CB - Concrete Block Seal   OB - 40-inch Omega Block Seal
HB - 40-inch Hybrid Omega block seal   # - Cribs
... - Roof Bolt Plates with Pie Pans and Belt Hangers Arranged Alternatively

20 psi

NIOSH - MSHA - WVOMHS&T Seal Testing
Test No. 1 Protocol -- LLEM test #501

ATTACHMENT 1

40-inch thick Omega Block Seal - 2001

The 40-inch thick Omega Block Seal was constructed at the Lake Lynn Laboratory, Experimental Mine, crosscut No. 3 in 2001 by Burrell Mining Products. NIOSH personnel assisted with construction, spotting material, mixing BlocBond and documentation.

- Average crosscut dimensions at seal location - 19-ft wide x 6-ft 9-in high.
- No hitching used on this seal.
- BlocBond applied approximately ¼-in thick to floor before starting first row.
- Approximately 264, 8-in x 16-in x 24-in Omega Blocks were used. Average weight of block was 45.7 lbs.
- Construction of the front course consisted of nine 8-in x 24-in x 16-in Omega Block plus one 8-in x 12-in x 16-in Omega Block. Construction of the back course consisted of twelve 8-in x 16-in x 24-in Omega Blocks plus one 8-in x 8-in x 24-in Omega Block. Eight courses thereafter were alternated to stagger joints.
- Final seal thickness was 40-inches plus coatings.
- Quikcrete BlocBond high strength, fiber-reinforced Surface Bonding Cement (average weight of 50 lb per bag) was applied as joint mortar and sealant on both sides. The mix of BlocBond consisted of 2 bags per approximately 4 gallons of water.
- BlocBond was applied approximately ¼-in thick on top and on all four sides of each block.
- The gap between the last course and the top of the seal was approximately 2.5 inches.
- Three rows of 1-in x 8-in x 10-foot hardwood boards were run lengthwise from rib to rib. One row of wood was placed in the middle of the seal and two rows of wood placed symmetrically on each side of the center piece with their respective edges flush with the inby and outby side of the seal. Each row of wood was wedged on about 1 foot centers. The gap between the wedges and the wood rows was filled with BlocBond.
- Both faces of the seal were coated with approximately ¼-in thick BlocBond.
- Construction of seal took approximately 9-1/2 hours.

NIOSH - MSHA- WVOMHS&T Seal Testing
Test No. 1 Protocol -- LLEM test #501

ATTACHMENT 2

Method of Construction for Hybrid 40-inch thick Omega Block Seal

- Clear loose material from the ribs, roof, and floor in the location of the seal and for a distance of at least 3 feet on each side of the seal's intended location. No hitching of the seal is required.
- Dampen entire crosscut (roof, ribs, and floor) with a fine water spray. If needed, re-dampen area where seal is to be constructed. Apply approximately a ¼-inch thick layer of dry BlocBond on the floor where the seal is to be constructed. Dampen top surface of BlocBond with a fine water spray to a wet but not "runny" condition.
- Lay the first course of Omega Block on the floor from rib to rib with approximately a ¼-inch gap in all vertical joints. Dampen the bottom of each Omega Block with a fine water spray prior to installation. The Omega Block should be laid without mortared joints and in a normal 40-inch thick pattern. If needed, cut block to fit.
- Mix BlocBond and water to a cement-like consistency according to manufacturer's recommendations. Vary consistency as needed.
- Pour mixture on top of the Omega Blocks and spread by gloved hand. Force BlocBond into vertical joints by gloved hand.
- An application of BlocBond is to be applied to the outby face of each row of blocks as they are constructed.
- For each subsequent row of blocks, repeat the previous 3 steps, but reverse the pattern of laying the Omega Blocks with mortar applied to the vertical joints with gloved hand on adjacent courses in order to stagger the vertical joints. If needed, cut block to fit from rib to rib. Continue additional courses of Omega Blocks until seal is within approximately 2.5 inches of the roof.
- One 1-in x 6-in x 16-ft board of rough cut lumber is to be placed on top of the seal and flush with the outby side of the seal. This board may be cut as needed so that there is approximately a 1-ft gap to each rib. Wedges are to be driven parallel to the board and skin-to-skin between the board and the roof.
- BlocBond is to be thrown along the top of seal to attempt to fill the gap between the top of the seal and the roof, from the outby board to near the center of the seal.
- A 1-in x 6-in x 16-ft board of rough cut lumber is to be placed on top of the seal near the center. This board may be cut as needed so that there is approximately a 1-ft gap to each rib. BlocBond is to be thrown along the top of seal to fill the gap between the top of the second board and the roof.
- A 1-in x 6-in x 16-ft board of rough cut lumber is to be placed on top of the seal and flush with the inby side of the seal. This board may be cut as needed so that there is approximately a 1-ft gap to each rib. Wedges are to be driven lengthwise and skin-to-skin between the board and the roof.

NIOSH - MSHA- WVOMHS&T Seal Testing
Test No. 1 Protocol -- LLEM test #501

- An application of an additional coat of BlocBond is to be applied to both faces of the seal from floor to roof and rib to rib with a thickness of ¼-inch.
- Dampen entire crosscut (roof, rib, and floor), including the base of the seal but excluding the faces of the seal, with a fine water spray once every 24 hours thereafter. No puddles or running water should be created.

NIOSH - MSHA- WVOMHS&T Seal Testing
Test No. 2 Protocol -- LLEM test #502

MSHA and WVOMHS&T are planning a series of tests at NIOSH's Lake Lynn Laboratory, Experimental Mine, to evaluate 40-inch thick Omega Block Seals. These tests are being performed as a direct result of the fatal explosion that occurred at the Sago Mine on January 2, 2006, where ten 40-inch thick Omega Block Seals were destroyed. Individual tests, within this series, will be designed to address the results of preceding tests, requiring a separate protocol for each test. This series of tests will assist investigators in determining why 40-inch thick Omega Block Seals failed at the Sago Mine.

Based on preliminary information, the following may have occurred at the Sago Mine:
1) Explosion pressures may have exceeded 20 psi;
2) The seals may not have been properly constructed;
3) Construction materials may have been substandard;
4) Cribs inby the seals may have contributed to the failure of the seals; and
5) A combination of any or all of the four items listed above.

Test No. 1 was completed on Saturday April 15. The pre-explosion inspection was conducted on Wednesday April 5. The post-explosion inspection was conducted on Wednesday April 19. In this test, a 40-inch thick Omega Block Seal was constructed on Thursday March 23 in the No. 2 crosscut as tested and approved in 2001. A second 40-inch Omega Block Seal was constructed on March 24 in the No. 3 crosscut with certain modifications. These seals were both subjected to a static explosion pressure in excess of 20 psi. Both seals passed the explosion and subsequent leakage test with only slight deformation.

Test No. 2 is a continuing baseline evaluation of the standard seals with an additional Omega Block seal constructed across the No. 3 entry. This is the entry in which the explosion was propagated.

Test No. 2:
As both of the seals passed Test No. 1, they will be used without any changes in Test No. 2. All of the damaged pie pans and belt hangers from Test No. 1 will be replaced. A 40-inch thick Omega Block Seal will be constructed in the No. 3 entry outby the No. 3 crosscut on May 18. It will be allowed to cure for 28 days. The same quantity of methane and coal dust used in Test No. 1 will be used in Test No. 2. The Test No. 2 explosion will be conducted on Thursday June 15. The pre-explosion inspection will be conducted on Tuesday June 13. The post-explosion inspection was conducted on Wednesday June 21.

Purpose:
This second explosion test will attempt to evaluate the results of both a static and dynamic pressure pulse from an explosion on the seals. The explosion pressure within the test area may result in pressure piling of an unknown magnitude. The amount of pressure is expected to be higher than in Test No. 1. This test method will allow NIOSH to evaluate

NIOSH - MSHA - WVOMHS&T Seal Testing
Test No. 2 Protocol -- LLEM test #502

the explosion pressure for the dynamic pressure pulse situation and to calibrate the system for future tests.

This test is not intended to duplicate the explosion and seal failure that occurred at the Sago Mine, and the results should not be interpreted as a replication of those events. This test is intended to provide a basis for future seal tests that may offer insight about the Sago Mine seal failures.

Method and Protocol:

The 40-inch thick Omega Block Seal in the No. 2 Crosscut was constructed on Thursday March 23 as tested and approved in 2001. This seal was built by Burrell Mining Products personnel. NIOSH personnel documented the construction techniques. The seal passed Test No. 1.

The hybrid 40-inch thick Omega Block Seal was constructed on Friday March 24 in the No. 3 Crosscut by NIOSH contractors under the direction of MSHA and WVOMHS&T personnel using construction techniques as the approved seal, with three exceptions. These exceptions include; applying unmixed mortar on the mine floor, not applying mortar directly to the vertical joints of the first course of blocks, and modifying the installation of wood planks and wedges between the last course of the Omega Blocks and the mine roof. The seal passed Test No. 1.

A 40-inch thick Omega Block Seal will be constructed on May 18 by NIOSH contractors under the direction of MSHA and WVOMHS&T in the No. 3 Entry outby the No. 3 Crosscut as tested and approved in 2001.

After a 28-day curing period, the seals will be subjected to an explosion using the same volume of methane and coal dust used in Test No. 1. The overpressure is expected to be larger than in Test No. 1

NIOSH - MSHA- WVOMHS&T Seal Testing
Test No. 2 Protocol -- LLEM test #502

Test No. 2

1) A Solid Concrete Block Seal has been installed in the 1st Crosscut.
2) Burrell Mining Products has installed a 40-inch thick Omega Block Seal in the 2nd Crosscut between the Nos. 2 and 3 Entries according to the approved 2001 methods.
3) A hybrid 40-inch thick Omega Block Seal has been installed in the 3rd Crosscut between the Nos. 2 and 3 Entries.
4) NIOSH contractors will install a 40-inch thick Omega Block Seal in the No. 3 entry outby the No. 3 Crosscut. (See Attachment 1)
5) Install belt hangers along the roof in the No. 3 Entry and roof bolt bearing plates with pie pans along the roof in the No. 3 Entry. Locate hangers and plates alternatively along the entry.
6) Record all pre-explosion parameters, including volume and percentage of methane.
7) Test with a pressure pulse 1 from an explosion using the same volume of methane and coal dust used in Test No. 1 28 days after completion.
8) Determine overpressures throughout the test area.
9) Record and map all post explosion results, including overpressures and flame length.

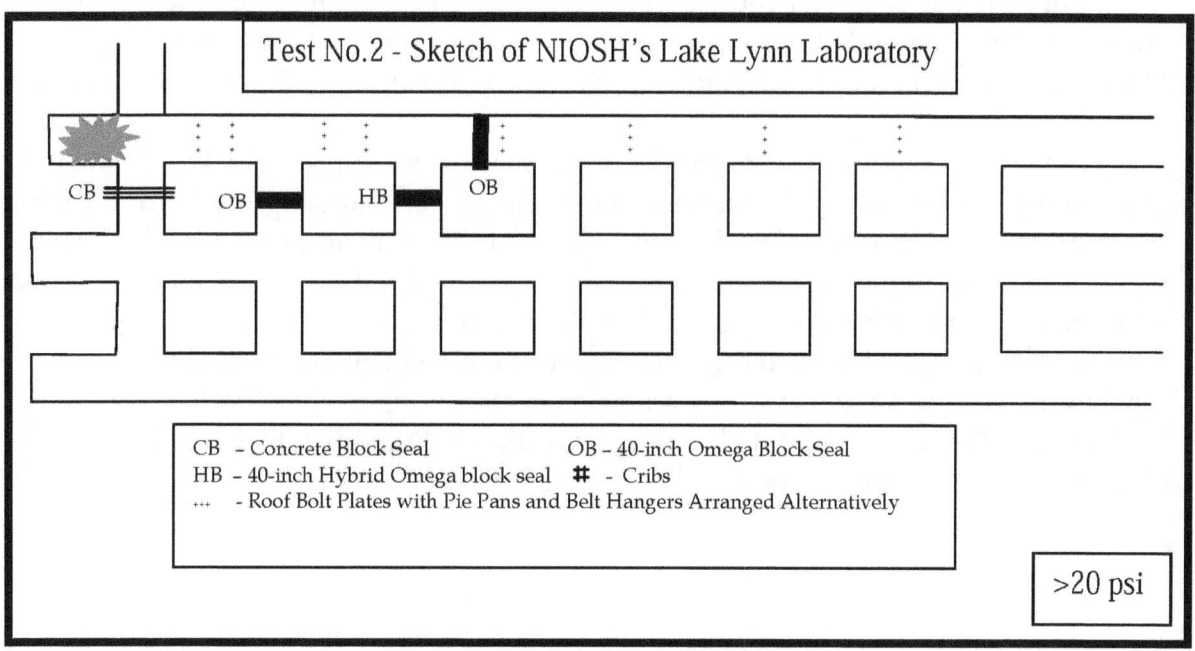

NIOSH - MSHA- WVOMHS&T Seal Testing
Test No. 2 Protocol -- LLEM test #502

ATTACHMENT 1

<u>40-inch thick Omega Block Seal - 2001</u>

The 40-inch thick Omega Block Seal was constructed at the Lake Lynn Laboratory, Experimental Mine, Crosscut No. 3 in 2001 by Burrell Mining Products. NIOSH personnel assisted with construction, spotting material, mixing BlocBond and documentation.

- Average entry dimensions at seal location - 19-ft wide x 6-ft 9-in high.
- No hitching used on this seal.
- BlocBond applied approximately ¼-in thick to floor before starting first row.
- Approximately 264, 8-in x 16-in x 24-in Omega Blocks were used. Average weight of block was 45.7 lbs.
- Construction of the front course consisted of nine 8-in x 24-in x 16-in Omega Block plus one 8-in x 12-in x 16-in Omega Block. Construction of the back course consisted of twelve 8-in x 16-in x 24-in Omega Blocks plus one 8-in x 8-in x 24-in Omega Block. Eight courses thereafter were alternated to stagger joints.
- Final seal thickness was 40-inches plus coatings.
- Quikcrete BlocBond high strength, fiber-reinforced Surface Bonding Cement (average weight of 50 lb per bag) was applied as joint mortar and sealant on both sides. The mix of BlocBond consisted of 2 bags per approximately 4 gallons of water.
- BlocBond was applied approximately ¼-in thick on top and on all four sides of each block.
- The gap between the last course and the top of the seal was approximately 2.5 inches.
- Three rows of 1-in x 8-in x 10-foot hardwood boards were run lengthwise from rib to rib. One row of wood was placed in the middle of the seal and two rows of wood placed symmetrically on each side of the center piece with their respective edges flush with the inby and outby side of the seal. Each row of wood was wedged on about 1 foot centers. The gap between the wedges and the wood rows was filled with BlocBond.
- Both faces of the seal were coated with approximately ¼-in thick BlocBond.
- Construction of seal took approximately 9-1/2 hours.

NIOSH - MSHA - WVOMHS&T Seal Testing
Test No. 3 Protocol -- LLEM test #503

MSHA and WVOMHS&T are involved in a series of tests at NIOSH's Lake Lynn Laboratory, Experimental Mine, to evaluate the effects of explosion pressures on the various seals, including the 40-inch thick Omega Block Seals. These tests are being performed as a direct result of the fatal explosion that occurred at the Sago Mine on January 2, 2006, where ten 40-inch thick Omega Block Seals were destroyed. Individual tests, within this series, will be designed to address the results of preceding tests, requiring a separate protocol for each test. Test No. 1 and Test No. 2 were completed on April 15 and June 15, 2006, respectively. Test No. 3 will be the next test in this series. This series of tests will assist investigators in determining why 40-inch thick Omega Block Seals failed at the Sago Mine.

Based on preliminary information, the following may have occurred at the Sago Mine:
1) Explosion pressures may have exceeded 20 psi;
2) The seals may not have been properly constructed;
3) Construction materials may have been substandard;
4) Cribs inby the seals may have contributed to the failure of the seals; and
5) A combination of any or all of the four items listed above.

Test No. 3:

Since the Solid Concrete Block Seal in Crosscut No. 1 and the Omega block seal in Crosscut No. 2 passed Test No. 1 and No. 2, they will be used without changes in Test No. 3. All of the damaged pie pans and belt hangers will be replaced. Two Sago 40-inch thick Omega Block Seals will be constructed as described in Attachment 1 in the No. 3 crosscut and in Drift C outby the No. 3 crosscut. This test is designed to evaluate the conditions and construction techniques of the seals at the Sago Mine. They will be allowed to cure for at least 28 days. The quantity of methane and coal dust used will subject the seals to an approximately 20 psi pressure pulse from an explosion.

Purpose:

This third explosion test will attempt to evaluate the results of both a static and a dynamic pressure pulse from an explosion on the seals. The explosion pressure within the test area is expected to be approximately 20 psi. This test is intended to help investigators evaluate the explosion and seal failure that occurred at the Sago Mine. Further tests are anticipated to complete this evaluation.

NIOSH - MSHA- WVOMHS&T Seal Testing
Test No. 3 Protocol -- LLEM test #503

Method and Protocol:

The Sago 40-inch thick Omega Block Seal will be constructed in the No. 3 Crosscut and in Drift C outby the No. 3 crosscut by NIOSH contractors under the direction of MSHA and WVOMHS&T personnel using the same construction techniques as used in the approved seal, with three exceptions. These exceptions include; applying unmixed mortar on the mine floor, not applying mortar directly to the vertical joints, and modifying the installation of wood planks and wedges between the last course of the Omega Blocks and the mine roof.

After at least a 28-day curing period, the seals will be subjected to an approximately 20 psi pressure pulse from an explosion.

Procedure:

1) A Solid Concrete Block Seal has been installed in the 1st Crosscut between Drift B and Drift C. This seal has been in place for each of the two preceding tests.

2) Burrell Mining Products has installed a 40-inch thick Omega Block Seal in the 2nd Crosscut between Drift B and Drift C according to the approved 2001 methods. This seal has been in place for each of the two preceding tests.

3) A new Sago 40-inch thick Omega Block Seal will be installed in the 3rd Crosscut between Drift B and Drift C and also in Drift C outby the No. 3 Crosscut by NIOSH contractors.

4) Two hollow-core 6-inch x 8-inch x 16-inch concrete block stoppings will be constructed as a part of Test No. 3. The stoppings will be dry stacked and coated on both sides with B-Bond sealant. The first stopping will be located across Drift C in the next pillar outby the Sago seal in Drift C, which is between crosscut Nos. 4 and 5. The second stopping will be located in Crosscut No. 3, between Drift A and Drift B.

5) Two wood cribs will be built approximately 5 feet inby the seal in Drift C and two wood cribs will be built approximately 5 feet outby the seal in Drift C.

6) Where necessary, install new belt hangers along the roof in Drift C and new roof bolt bearing plates with pie pans along the roof in Drift C. Locate hangers and plates alternatively along the entry.

7) Record all pre-explosion parameters, including volume and percentage of methane.

NIOSH - MSHA- WVOMHS&T Seal Testing
Test No. 3 Protocol -- LLEM test #503

8) Test with a pressure pulse of approximately 20 psi, at least 28 days after completion of the final seal.

9) Determine overpressures throughout the test area.

10) Record and map all post explosion results, including overpressures and flame length.

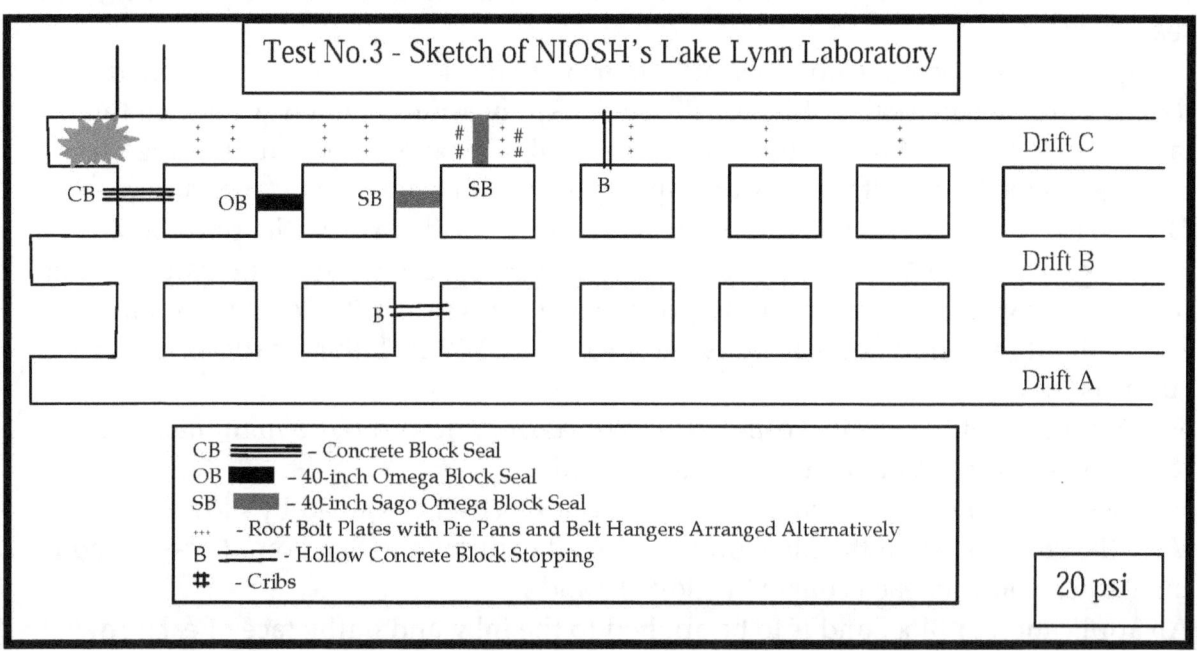

NIOSH - MSHA- WVOMHS&T Seal Testing
Test No. 3 Protocol -- LLEM test #503

ATTACHMENT 1

Method of Construction for the Sago 40-inch thick Omega Block Seal
Test No. 3

- Clear loose material from the ribs, roof, and floor in the location of the seal and for a distance of at least 3 feet on each side of the seal's intended location. No hitching of the seal is required.
- Dampen entire crosscut (roof, ribs, and floor) with a fine water spray. If needed, re-dampen area where seal is to be constructed. Apply approximately a 1.5-inch thick layer of dry BlocBond on the floor where the seal is to be constructed. Dampen top surface of BlocBond with a fine water spray to a wet but not "runny" condition.
- Lay the first course of Omega Block on the floor from rib to rib, with approximately a ¼-inch gap in all vertical joints. Dampen the bottom of each of the Omega Blocks on the first row with a fine water spray prior to installation. The Omega Block should be laid without mortared vertical joints and in a normal 40-inch thick pattern. If needed, cut block to fit.
- Mix BlocBond and water to a cement-like consistency according to manufacturer's recommendations. Vary consistency as needed.
- Apply mixture on top of each course of Omega Blocks and spread by gloved hand. Vary the consistency of the BlocBond as needed to allow the BlocBond to be forced into the vertical joints of each course by gloved hand.
- An application of BlocBond is to be applied to the inby and outby face of each row of blocks as they are constructed.
- For each subsequent row of blocks, repeat the previous 3 steps, but reverse the pattern of laying the Omega Blocks on adjacent courses in order to stagger the vertical joints. The Omega Blocks should be laid with a ¼-inch vertical joint gap and those joints should not be mortared. If needed, cut block to fit from rib to rib. Continue additional courses of Omega Blocks until seal is within approximately 2.5 inches of the roof.
- One-inch x six-inch boards of rough cut lumber are to be placed on top of the seal near the center. These boards may be cut as needed so that they span the distance from rib to rib to the extent practicable. An attempt will be made to drive wedges parallel to the board and skin-to-skin between the board and the roof to the extent practicable.
- BlocBond is to be thrown along the top of seal to fill any remaining gap between the top of the center board and the roof and the area between the center board and the location of the inby and outby board.
- One-inch x six-inch boards of rough cut lumber are to be placed on top of the seal and flush with the outby side of the seal. These boards may be cut as needed so that they

NIOSH - MSHA- WVOMHS&T Seal Testing
Test No. 3 Protocol -- LLEM test #503

span the distance from rib to rib. Wedges are to be driven parallel to the board and skin-to-skin between the board and the roof.
- One-inch x six-inch boards of rough cut lumber are to be placed on top of the seal and flush with the inby side of the seal. These boards may be cut as needed so that they span the distance from rib to rib. Wedges are to be driven parallel to the board and skin-to-skin between the board and the roof.
- BlocBond is to be thrown along the top of seal to fill any remaining gap between the top of the boards and the roof
- An application of an additional coat of BlocBond is to be applied to both faces of the seal from floor to roof and rib to rib with a thickness of ¼-inch.
- Dampen entire crosscut (roof, rib, and floor), including the base of the seal but excluding the faces of the seal, with a fine water spray once every 24 hours thereafter. No puddles or running water should be created.

NIOSH - MSHA - WVOMHS&T Seal Testing
Test No. 4 Protocol -- LLEM test #504

MSHA and WVOMHS&T are involved in a series of tests at NIOSH's Lake Lynn Laboratory, Experimental Mine, to evaluate the effects of explosion pressures on the various seals, including the 40-inch thick Omega Block Seals. These tests are being performed as a direct result of the fatal explosion that occurred at the Sago Mine on January 2, 2006, where ten 40-inch thick Omega Block Seals were destroyed. Individual tests, within this series, will be designed to address the results of preceding tests, requiring a separate protocol for each test. Tests No. 1, No. 2 and 3 were completed on April 15, June 15, and August 4, 2006, respectively. However, Test No. 3 did not expose the seals to the expected 20 psi explosion pressure. Test No. 4 will be the next test in this series. This series of tests will assist investigators in determining why 40-inch thick Omega Block Seals failed at the Sago Mine.

Based on preliminary information, the following may have occurred at the Sago Mine:
1) Explosion pressures may have exceeded 20 psi;
2) The seals may not have been properly constructed;
3) Construction materials may have been substandard;
4) Cribs inby the seals may have contributed to the failure of the seals; and
5) A combination of any or all of the four items listed above.

Test No. 4:

Since the Solid Concrete Block Seal in Crosscut No. 1 and the Omega block seal in Crosscut No. 2 passed Tests No. 1, 2 and 3, they will be used without changes in Test No. 4. Also, since the two Sago 40-inch thick Omega Block Seals passed Test 3, they will be used without change in Test No. 4. This test is designed to evaluate the conditions and construction techniques of the seals at the Sago Mine. The quantity of methane and coal dust used will subject the seals to an approximate 20 psi pressure pulse from an explosion.

Purpose:

This fourth explosion test will attempt to evaluate the results of both a static and a dynamic pressure pulse from an explosion on the seals. The explosion pressure within the test area is expected to be approximately 20 psi. This test is intended to help investigators evaluate the explosion and seal failure that occurred at the Sago Mine. Further tests are anticipated to complete this evaluation.

NIOSH - MSHA - WVOMHS&T Seal Testing
Test No. 4 Protocol -- LLEM test #504

Method and Protocol:

Existing constructed seals will be used and no seals will need to be built.
The seals will be subjected to an approximate 20 psi pressure pulse from an explosion.

Procedure:

1) A Solid Concrete Block Seal has been installed in the 1st Crosscut between Drift B and Drift C. This seal has been in place for each of the three preceding tests.

2) Burrell Mining Products has installed a 40-inch thick Omega Block Seal in the 2nd Crosscut between Drift B and Drift C according to the approved 2001 methods. This seal has been in place for each of the three preceding tests.

3) A Sago 40-inch thick Omega Block Seal has been installed in the 3rd Crosscut between Drift B and Drift C and also in Drift C outby the No. 3 Crosscut. These seals were in place for Test No. 3.

4) Two hollow-core 6-inch x 8-inch x 16-inch concrete block stoppings were constructed as a part of Test No. 3. The stoppings were dry stacked and coated on both sides with B-Bond sealant. The first stopping is located across Drift C in the next pillar outby the Sago seal in Drift C, which is between crosscut Nos. 4 and 5. The second stopping is located in Crosscut No. 3, between Drift A and Drift B.

5) Wood cribs built as part of Test No. 3 will remain in place. Two cribs are approximately 5 feet inby the seal in Drift C and two cribs are approximately 5 feet outby the seal in Drift C.

6) Record all pre-explosion parameters, including volume and percentage of methane.

7) Test with a pressure pulse of approximately 20 psi.

8) Determine overpressures throughout the test area.

9) Record and map all post explosion results, including overpressures and flame length.

NIOSH - MSHA- WVOMHS&T Seal Testing
Test No. 4 Protocol -- LLEM test #504

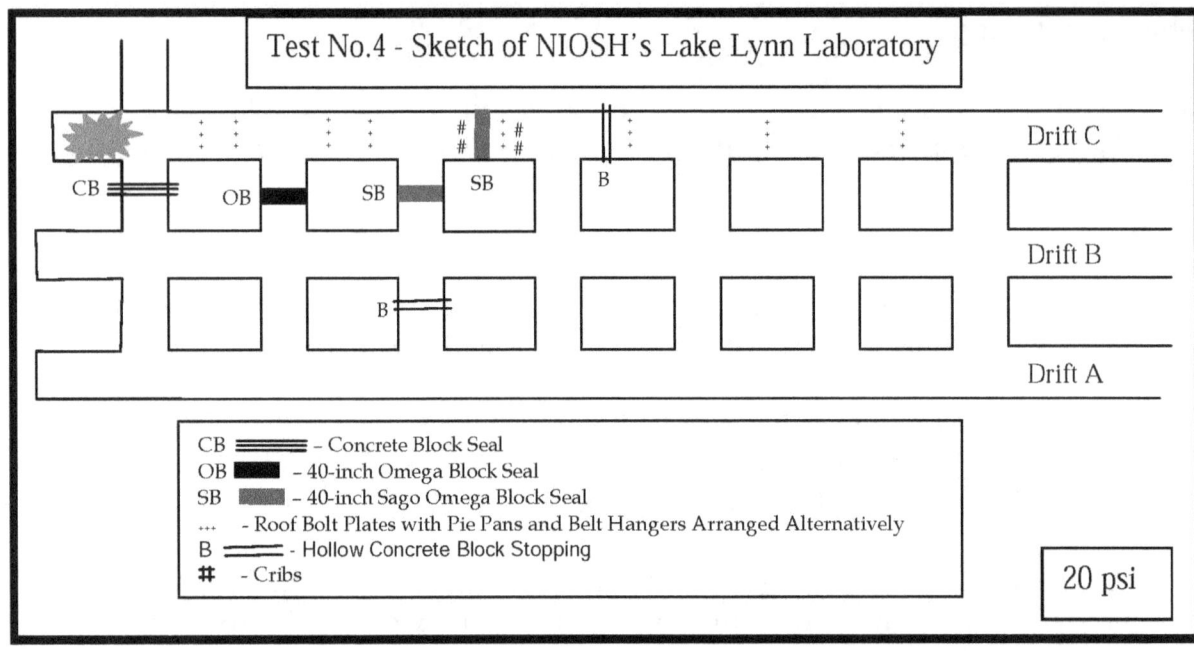

NIOSH - MSHA- WVOMHS&T Seal Testing
Test No. 5 Protocol -- LLEM test #505

MSHA and WVOMHS&T are involved in a series of tests at NIOSH's Lake Lynn Laboratory, Experimental Mine, to evaluate the effects of explosion pressures on the various seals, including the 40-inch thick Omega Block Seals. These tests are being performed as a direct result of the fatal explosion that occurred at the Sago Mine on January 2, 2006, where ten 40-inch thick Omega Block Seals were destroyed. Individual tests, within this series, will be designed to address the results of preceding tests, requiring a separate protocol for each test. Tests Nos. 1, 2, 3 and 4 were completed on April 15, June 15, August 4, and August 16, 2006, respectively. Test No. 5 will be the next test in this series. This series of tests will assist investigators in determining why 40-inch thick Omega Block Seals failed at the Sago Mine.

Test No. 5:

Since the Solid Concrete Block Seal in Crosscut No. 1 and the Omega block seal in Crosscut No. 2 passed Tests Nos. 1, 2, 3 and 4, they will be used without changes in Test No. 5. Also, since the two Sago 40-inch thick Omega Block Seals passed Tests 3 and 4, they will be used without change in Test No. 5. This test is designed to evaluate the conditions and construction techniques of the seals at the Sago Mine. The quantity of methane and coal dust used will subject the seals to an approximate 50 psi pressure pulse from an explosion.

Purpose:

This fifth explosion test will attempt to evaluate the results of both a static and a dynamic pressure pulse from an explosion on the seals. The explosion pressure within the test area is expected to be approximately 50 psi. This test is intended to help investigators evaluate the explosion and seal failure that occurred at the Sago Mine. Further tests are anticipated to complete this evaluation.

Method and Protocol:

Existing constructed seals will be used and no seals will need to be built.
The seals will be subjected to an approximate 50 psi pressure pulse from an explosion.

Procedure:

1) A Solid Concrete Block Seal has been installed in the 1st Crosscut between Drift B and Drift C. This seal has been in place for each of the four preceding tests.

2) Burrell Mining Products has installed a 40-inch thick Omega Block Seal in the 2nd Crosscut between Drift B and Drift C according to the approved 2001 methods. This seal has been in place for each of the four preceding tests.

NIOSH - MSHA- WVOMHS&T Seal Testing
Test No. 5 Protocol -- LLEM test #505

3) A Sago 40-inch thick Omega Block Seal has been installed in the 3rd Crosscut between Drift B and Drift C and also in Drift C outby the No. 3 Crosscut. These seals were in place for Test Nos. 3 and 4.

4) Two hollow-core 6-inch x 8-inch x 16-inch concrete block stoppings were constructed as a part of Test Nos. 3 and 4. The stoppings were dry stacked and coated on both sides with B-Bond sealant. The first stopping is located across Drift C in the next pillar outby the Sago seal in Drift C, which is between crosscut Nos. 4 and 5. The second stopping is located in Crosscut No. 3, between Drift A and Drift B.

5) Wood cribs built as part of Test Nos. 3 and 4 will remain in place. Two cribs are approximately 5 feet inby the seal in Drift C and two cribs are approximately 5 feet outby the seal in Drift C.

6) Record all pre-explosion parameters, including volume and percentage of methane.

7) Test with a pressure pulse of approximately 50 psi.

8) Determine overpressures throughout the test area.

9) Record and map all post explosion results, including overpressures and flame length.

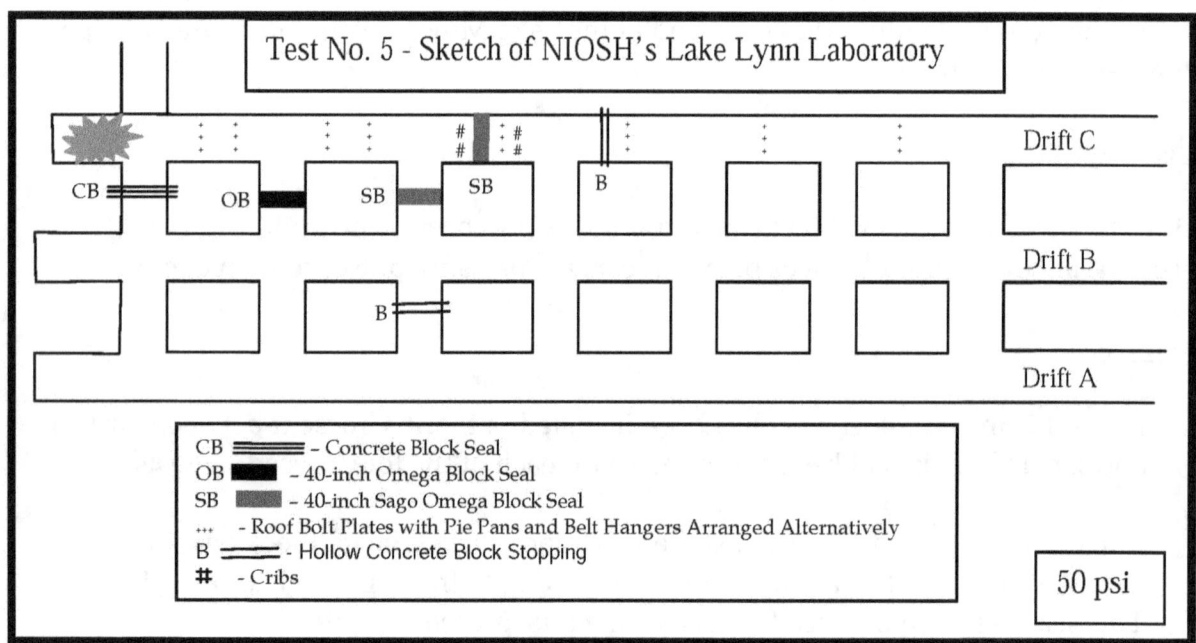

NIOSH - MSHA- WVOMHS&T Seal Testing
Test No. 6 Protocol -- LLEM test #506

MSHA and WVOMHS&T are involved in a series of tests at NIOSH's Lake Lynn Laboratory, Experimental Mine, to evaluate the effects of explosion pressures on the various seals, including the 40-inch thick Omega Block Seals. These tests are being performed as a direct result of the fatal explosion that occurred at the Sago Mine on January 2, 2006, where ten 40-inch thick Omega Block Seals were destroyed. Individual tests, within this series, have been designed to address the results of preceding tests, requiring a separate protocol for each test. Tests Nos. 1, 2, 3, 4 and 5 were completed on April 15, June 15, August 4, August 16 and August 23, respectively. Test No. 6 will be the next test in this series. This series of tests will assist investigators in determining why 40-inch thick Omega Block Seals failed at the Sago Mine.

Test No. 6:

Since the Solid Concrete Block Seal in Crosscut No. 1 and the Omega Block Seal in Crosscut No. 2 passed Tests Nos. 1, 2, 3, 4 and 5, they will be used without changes in Test No. 6. The two Sago 40-inch thick Omega Block Seals in Drift C and in Crosscut No. 3 were destroyed in Test 5. Therefore, a Solid Concrete Block Seal will be build in Crosscut No. 3 and a Sago 40-inch thick Omega Block Seal using the blocks supplied from the Sago Mine supplemented with additional Omega Blocks obtained from Burrell Mining Products, if needed, will be built in Drift C. This test is designed to evaluate the conditions and construction techniques of the seals at the Sago Mine. After at least a 28-day curing period, the seals will be subjected to an approximately 90 psi pressure pulse from an explosion.

Purpose:

This sixth explosion test will attempt to evaluate the results of a pressure pulse from an explosion on the seals. The explosion pressure on the Sago Seal is expected to be approximately 90 psi. This test is intended to help investigators evaluate the explosion and seal failure that occurred at the Sago Mine.

Method and Protocol:

A Solid Concrete Block Seal will be build in Crosscut No. 3 and a Sago 40-inch thick Omega Block Seal using the blocks supplied from the Sago Mine will be built in Drift C supplemented with additional Omega Blocks obtained from Burrell Mining Products, if needed. The seals will be subjected to an approximate 90 psi pressure pulse from an explosion.

NIOSH - MSHA- WVOMHS&T Seal Testing
Test No. 6 Protocol -- LLEM test #506

Procedure:

1) A Solid Concrete Block Seal has been installed in the 1st Crosscut between Drift B and Drift C. This seal has been in place for each of the five preceding tests.

2) Burrell Mining Products has installed a 40-inch thick Omega Block Seal in the 2nd Crosscut between Drift B and Drift C according to the approved 2001 methods. This seal has been in place for each of the five preceding tests.

3) A Solid Concrete Block Seal will be built in the 3rd Crosscut between Drift B and Drift C.

4) A Sago 40-inch thick Omega Block Seal will be installed in Drift C outby the No. 3 Crosscut. The protocol is outlined in Appendix 1.

5) A hollow-core 6-inch x 8-inch x 16-inch concrete block stopping will be constructed as a part of Test 6. The stopping will be dry stacked and coated on both sides with B-Bond sealant. The stopping will be located across Drift C in the next pillar outby the Sago seal in Drift C, which is between crosscut Nos. 4 and 5.

6) Two wood cribs will be built approximately 5 feet inby the seal in Drift C and two cribs are approximately 5 feet outby the seal in Drift C.

7) Record all pre-explosion parameters, including volume and percentage of methane.

9) Test with a pressure pulse of approximately 90 psi.

9) Determine overpressures throughout the test area.

10) Record and map all post explosion results, including overpressures and flame length.

NIOSH - MSHA- WVOMHS&T Seal Testing
Test No. 6 Protocol -- LLEM test #506

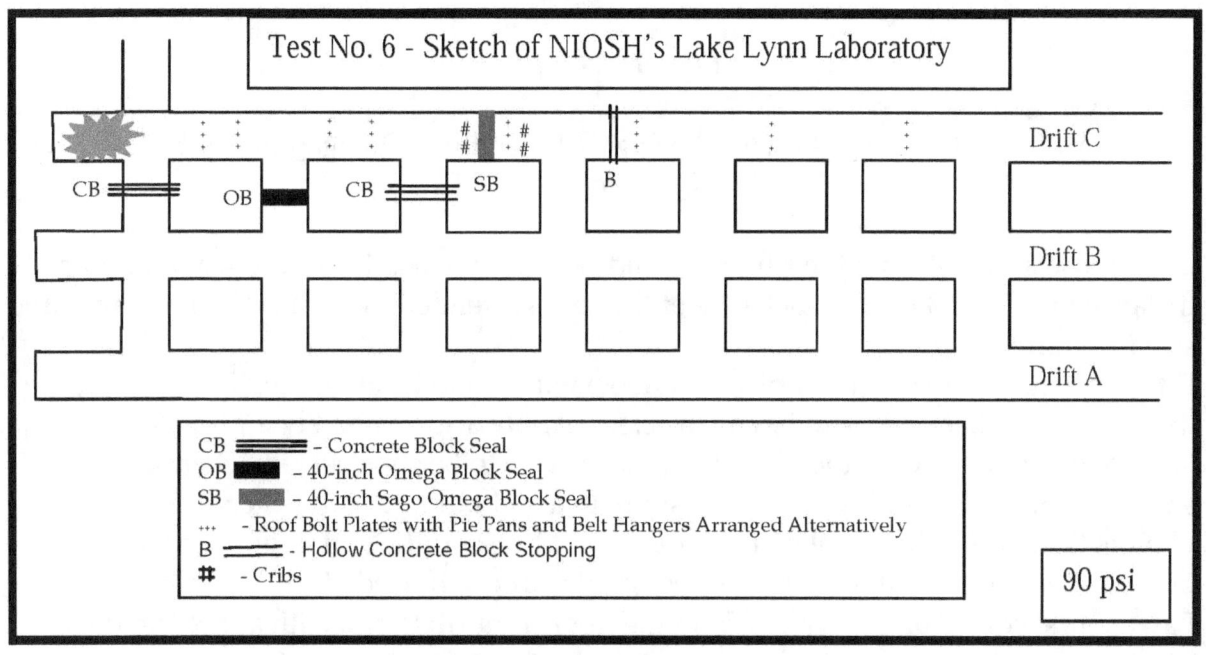

NIOSH - MSHA- WVOMHS&T Seal Testing
Test No. 6 Protocol -- LLEM test #506

ATTACHMENT 1

Method of Construction for the Sago 40-inch thick Omega Block Seal
Test No. 6

- Clear loose material from the ribs, roof, and floor in the location of the seal and for a distance of at least 3 feet on each side of the seal's intended location. No hitching of the seal is required.
- Dampen entire crosscut (roof, ribs, and floor) with a fine water spray. If needed, re-dampen area where seal is to be constructed. Apply approximately a 1.5-inch thick layer of dry BlocBond on the floor where the seal is to be constructed. Dampen top surface of BlocBond with a fine water spray to a wet but not "runny" condition.
- Use the Omega Blocks supplied from the Sago Mine supplemented with additional Omega Blocks obtained from Burrell Mining Products, if needed.
- Lay the first course of Omega Block on the floor from rib to rib, with approximately a ¼-inch gap in all vertical joints. Dampen the bottom of each of the Omega Blocks on the first row with a fine water spray prior to installation. The Omega Block should be laid without mortared vertical joints and in a normal 40-inch thick pattern. If needed, cut block to fit.
- Mix BlocBond and water to a cement-like consistency according to manufacturer's recommendations. Vary consistency as needed.
- Apply mixture on top of each course of Omega Blocks and spread by gloved hand. Vary the consistency of the Blocbond as needed to allow the BlocBond to be forced into the vertical joints of each course by gloved hand.
- An application of BlocBond is to be applied to the inby and outby face of each row of blocks as they are constructed.
- For each subsequent row of blocks, repeat the previous 3 steps, but reverse the pattern of laying the Omega Blocks on adjacent courses in order to stagger the vertical joints. The Omega Blocks should be laid with a ¼-inch vertical joint gap and those joints should not be mortared. If needed, cut block to fit from rib to rib. Continue additional courses of Omega Blocks until seal is within approximately 2.5 inches of the roof.
- One-inch x six-inch boards of rough cut lumber are to be placed on top of the seal near the center. These boards may be cut as needed so that they span the distance from rib to rib to the extent practicable. An attempt will be made to drive wedges parallel to the board and skin-to-skin between the board and the roof to the extent practicable.
- BlocBond is to be thrown along the top of seal to fill any remaining gap between the top of the center board and the roof and the area between the center board and the location of the inby and outby board.

NIOSH - MSHA- WVOMHS&T Seal Testing
Test No. 6 Protocol -- LLEM test #506

- One-inch x six-inch boards of rough cut lumber are to be placed on top of the seal and flush with the outby side of the seal. These boards may be cut as needed so that they span the distance from rib to rib. Wedges are to be driven parallel to the board and skin-to-skin between the board and the roof.
- One-inch x six-inch boards of rough cut lumber are to be placed on top of the seal and flush with the inby side of the seal. These boards may be cut as needed so that they span the distance from rib to rib. Wedges are to be driven parallel to the board and skin-to-skin between the board and the roof.
- BlocBond is to be thrown along the top of seal to fill any remaining gap between the top of the boards and the roof
- An application of an additional coat of BlocBond is to be applied to both faces of the seal from floor to roof and rib to rib with a thickness of ¼-inch.
- Dampen entire crosscut (roof, rib, and floor), including the base of the seal but excluding the faces of the seal, with a fine water spray once every 24 hours thereafter. No puddles or running water should be created.

APPENDIX B.—SEAL CONSTRUCTION DESCRIPTIONS

After the Omega block seal evaluations in the LLEM during 2006, the following seal construction descriptions were completed in February 2007.[1]

1. Standard 40-in Omega Block Seal, 2001 Design, in Crosscut 2
Construction Date: March 23, 2006
NIOSH–MSHA–WVOMHS&T Seal Testing – Test No. 1 Protocol

On March 23, 2006, a nominal 40-in-thick Omega block seal was constructed by personnel from Burrell Mining Products International in crosscut 2 between B- and C-drifts in the LLEM. It was constructed of Omega 384 low-density blocks manufactured by Burrell Mining Products. This seal was based on the 2001 design. When the coating thickness on the faces of the seal and the mortar thickness are included, the total seal thickness was about 41 in.

- Average crosscut dimensions at seal location: 18.8 ft wide by 6.7 ft high.
- No hitching was used on this seal.
- The seal was constructed approximately 4 ft into the crosscut (as measured from the C-drift side) on a small concrete foundation that tapered from 0 to 3 in thick on top of an 8-in-thick reinforced concrete floor designed to assist in the leveling of the first course of blocks.
- Quikrete BlocBond (1225-51) was mixed with water and applied approximately 0.25–0.5 in thick to the floor (concrete foundation) as the first course was being laid (Figure B-1).
- Construction of the first row (front course, C-drift side) consisted of nine 8-in by 24-in by 16-in Omega blocks (24-in block dimension parallel with C-drift) and one Omega block cut to fit. The gap between blocks was approximately 0.25–0.5 in. Construction of the first row (back course, B-drift side) consisted of fourteen 8-in by 16-in by 24-in Omega blocks (16-in block dimension parallel with B-drift) and one Omega block cut to fit. The gap between blocks was 0.25–0.5 in, and the gap between the front and back course varied approximately 0.25–0.5 in. The blocks were oriented so that the course height was about 8 in. Full Omega blocks were laid against each rib, and the cut blocks were fitted inside the row.
- Using shovels and gloved hands, BlocBond was applied to all contact surfaces of the Omega blocks. BlocBond was used to fill the gaps (Figure B-2) between the Omega blocks and gaps between the Omega blocks and ribs. As each Omega block course was completed, BlocBond was also coated on the exposed B- and C-drift sides of the blocks (Figure B-3).
- The remaining eight full-block courses were installed in a manner similar to the first course, except the blocks in each subsequent course were alternated to stagger the block joints. Figure B-4 is a schematic to illustrate alternating courses and stagger block joints.

---

[1]The seal construction descriptions in this appendix were written by Cynthia A. Hollerich, Eric S. Weiss, and Michael J. Sapko (retired) of the NIOSH Pittsburgh Research Laboratory.

- On the 10th and final block course, each of the blocks was cut and mortared into place to leave an approximately 2-in gap between the blocks and the mine roof (Figures B-5 and B-6).
- Approximately 233 new Omega blocks were used in the construction of this seal. These blocks were manufactured in the Garards Fort, PA, plant.
- Three rows of 1-in-thick, 8-in-wide, 10-ft-long hardwood rough-cut boards were run lengthwise (rib to rib) on the top course of Omega blocks. One row of wood was placed across the middle of the seal, one row of wood was placed flush with the edge of the block on the C-drift side, and another flush with the edge of the block on the B-drift side (Figure B-7). The rows of wood extended the entire length of the seal (rib to rib).
- To complete the seal closure to the mine roof, the cut Omega blocks on the B-drift side of the seal were first installed (24-in block dimension parallel with C-drift). A 10-ft-long rough-cut board was then coated on top with BlocBond, placed on the Omega block course tightly against the one rib and flush with the seal face, and then tightly wedged in place. A second board was installed in an identical manner across the remainder of the seal width on the B-drift side. All gaps between the wedges were filled with BlocBond, and all of the wood surfaces were coated with BlocBond. Work then resumed on the C-drift side of the seal where the last full Omega block front course (ninth course) was intentionally not installed to provide additional working space for installing the middle board across the center of the seal. Before installing this middle board, small Omega block pieces were first cut (approximately 8 in by 24 in by 6.5 in) and mortared in place to within about 2 in of the mine roof to complete the partial block course across the middle section of the seal. The rough-cut board for the middle section of the seal was then installed in a manner similar to the board on the B-drift side. Finally, the last full Omega block course (9th course) and the partial block top course (10th course) on the C-drift side were mortared into place. The rough-cut boards were then installed in an identical manner as on the B-drift side and coated with BlocBond (Figures B-8 and B-9).
- Using gloved hands, an approximately 0.25-in-thick coating of BlocBond was then applied to both faces of the seal.
- Construction of the seal took approximately 8.25 hr (41.25 worker-hr). This did not include the time required to spot the construction materials to the site.
- Ninety-three bags of BlocBond (1225-51) high-strength, fiber-reinforced surface bonding cement (50-lb average bag weight) were required to complete this seal, including that used as joint mortar and face sealant (approximately two bags of BlocBond were required to coat each face of the seal). Not including water weight, the total weight of BlocBond was approximately 4,650 lb. In general, each batch of mortar and sealant used for this seal consisted of two bags of BlocBond mixed with approximately 4.5–5.0 gal of water. This calculates to approximately 209–233 gal of water used with the 93 bags of BlocBond required for constructing the seal.

Figure B-1.—BlocBond applied to concrete mine floor.

Figure B-2.—BlocBond being applied to gaps between the Omega blocks.

Figure B-3.—As each Omega block course was completed, BlocBond was also coated on the exposed B- and C-drift sides of the blocks.

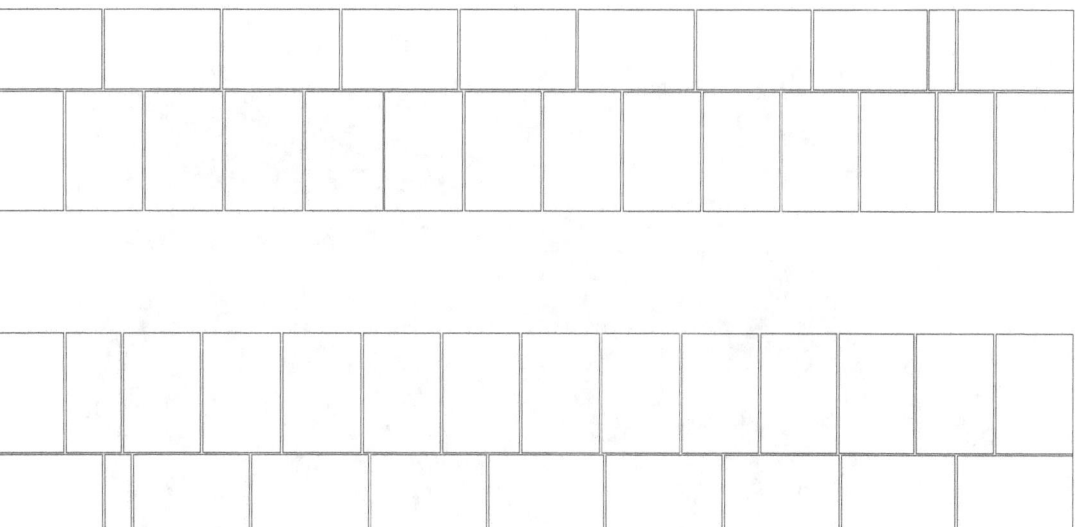

Figure B-4.—Schematic illustrating alternating courses and staggered block joints.

Figure B-5.—Cut blocks on top course to provide 2-in gap to roof.

Figure B-6.—View of top portion of seal from C-drift side after B-drift side was completed. The last two block courses and middle and inby crossboards were installed from the C-drift side to complete the seal.

Figure B-7.—One row of wood placed flush with the edge of the blocks on the B-drift side.

Figure B-8.—View from the C-drift side showing the rough-cut lumber wedged between the mine roof and blocks across the center of the seal.

Figure B-9.—Closeup of the wood wedges, installed between the mine roof and the rough-cut board, used to tighten the Omega block seal.

2. Hybrid 40-in Omega Block Seal Design in Crosscut 3
Construction Date: March 24, 2006
NIOSH–MSHA–WVOMHS&T Seal Testing – Test No. 1 Protocol

On March 24, 2006, a nominal 40-in-thick hybrid Omega block seal was constructed by personnel from Ki Corp. (a NIOSH contractor) in crosscut 3 between B- and C-drifts in the LLEM. It was constructed of Omega 384 low-density blocks manufactured by Burrell Mining Products. When the coating thickness on the faces of the seal and the mortar thickness are included, the total seal thickness was about 41 in.

- Average crosscut dimensions at seal location: 18.8 ft wide by 6.75 ft high.
- No hitching was used on this seal.
- The seal was constructed approximately 6–7 ft into the crosscut (as measured from the C-drift side) on a small concrete foundation that tapered from 0 to 3 in thick on top of an 8-in-thick reinforced concrete floor designed to assist in the leveling of the first course of blocks.
- The entire crosscut (roof, ribs, and floor) was dampened with a fine water spray just before starting the construction. An approximately 0.25-in-thick dry layer of Quikrete BlocBond (1225-51) was applied by hand on the floor (concrete foundation) where the seal was to be constructed (Figure B-10). This dry layer of BlocBond was then dampened with a fine water spray to a wet, but not runny condition (Figure B-11).
- Construction of the first row (front course, C-drift side) consisted of thirteen 8-in by 16-in by 24-in Omega blocks (16-in block dimension parallel with C-drift) and one Omega block cut to fit. Construction of the first row (back course, B-drift side) consisted of nine 8-in by 16-in by 24-in Omega blocks (24-in block dimension parallel with B-drift) and one Omega block cut to fit. There were irregular gaps (from 0.25 to 0.5 in depending on cut of block) left between all blocks. Using a fine spray from a water hose, the Omega blocks were dampened on the underside before positioning them onto the dampened layer of BlocBond on the floor. No BlocBond was applied to the block joints before positioning the blocks onto the floor (Figure B-12). The entire first course was laid in the same manner.
- Using shovels and gloved hands, BlocBond mixed with water was then applied to all accessible surfaces of this first course of Omega blocks (Figure B-13). BlocBond was used to fill the (vertical) gaps between the Omega blocks as completely as possible by gloved hand and also the gaps between the Omega blocks and ribs. As each Omega block course was completed, BlocBond was also coated on the exposed B- and C-drift sides of the blocks and forced into the outside vertical joints and gaps as completely as possible.
- The remaining eight Omega full-block courses were installed in a manner similar to the first course except BlocBond was applied by gloved hand to the block joints before placement and the blocks in each subsequent course were alternated to stagger the block joints.
- On the 10th and final course, each block was cut and mortared into place to leave an approximately 2.5-in gap between these blocks and the mine roof. The top course of blocks for the entire seal was mortared into place before installing any of the boards and wedges.

- Approximately 215 Omega blocks were used to construct this seal. These blocks were manufactured at the Garards Fort, PA, plant.
- Three rows of 1-in-thick, 6-in-wide, 16-ft-long hardwood rough-cut boards were run lengthwise and centered (between ribs) on the top course of Omega blocks. One row of wood was placed across the middle section of the seal, and one row of wood was placed on the flush end on each side of the seal. These rows of wood extended across the center of the seal, leaving an approximately 18-in gap toward each rib.
- One 1-in-thick by 6-in-wide by 16-ft rough-cut board was placed on top of the seal and flush with the outby (B-drift) side of the seal. There was an approximately 18-in gap between the centered board and each rib. Wedges were driven parallel to the board between the board and the roof. BlocBond was then thrown along the top of the seal and spread by gloved hand to fill any gaps (between wedges and the 18-in-long by 2.5-in-high gap between the end of the board and each rib) and to cover all exposed wood. Using gloved hands, an approximately 0.25-in-thick coating of BlocBond was then applied to the B-drift side of the seal. Working from the C-drift side, BlocBond was thrown by gloved hand along the top of the seal (between the top block course and mine roof) to fill the gap between the board on the B-drift side and the center of the seal. Then one 1-in-thick by 6-in-wide by 16-ft-long rough-cut board was pushed into place across the top of the seal near the center. An approximately 18-in gap remained between the ends of the centered board and each rib. BlocBond was again thrown with gloved hands along the top of the seal to fill any gaps between the wedges and middle board and the mine roof. Additional BlocBond was then thrown by gloved hand to fill the 2.5-in gap between the top block course and the mine roof from the middle board to where the adjacent board, which was flush-mounted on the C-drift side, will be positioned. Once this gap was filled as completely as possible with BlocBond, a final 1-in-thick by 6-in-wide by 16-ft-long rough-cut board was placed across the top of the seal and flush with the inby (C-drift) side of the seal. (No attempts were made to pack the BlocBond across the top of the seal using any tools or implements other than throwing the material by gloved hand.) As with the two other boards, an 18-in gap remained between the ends of this centered board and each rib. Wedges were driven parallel to the board and skin to skin between the board and the roof. BlocBond was again thrown along top of the seal to fill any gaps (between the wedges and the 18-in-long by 2.5-in-high gap between the end of the board and each rib) and to cover all exposed wood. Using gloved hands, an approximately 0.25-in-thick coating of BlocBond was then applied to the C-drift side of the seal.
- Ninety-two bags of BlocBond high-strength fiber-reinforced surface bonding cement (50-lb average bag weight) were required to complete this seal, including that used as joint mortar and face sealant (approximately two bags of BlocBond were required to coat each face of the seal). Not including water weight, the total weight of BlocBond was approximately 4,600 lb. In general, each batch of mortar and sealant used for this seal consisted of two bags of BlocBond mixed with approximately 4.25–4.50 gal of water. This calculates to approximately 195–207 gal of water used with the 92 bags of BlocBond required for constructing this seal.
- Construction of the seal took approximately 8 hr (32 worker-hr). This does not include the time required to spot the construction materials to the site.

Figure B-10.—Dry BlocBond being spread to a 0.5-in-thick layer on the dampened concrete floor in X-3.

Figure B-11.—Wetting the dry layer of BlocBond before positioning the Omega blocks.

Figure B-12.—Positioning the first course of Omega blocks.

Figure B-13.—Applying BlocBond to the top of the first course. The BlocBond is forced into the vertical joints using only gloved hands.

3. 40-in Omega Block Seal, 2001 Design, at 320 ft Outby C-Drift Face
Construction Date: May 18, 2006
NIOSH–MSHA–WVOMHS&T Seal Testing – Test No. 2 Protocol

On May 18, 2006, a nominal 40-in-thick Omega block seal (based on the design that was tested and approved in 2001) was constructed by personnel from Ki Corp. (a NIOSH contractor) in C-drift between crosscuts 3 and 4 (~C-320) in the LLEM. This seal was based on the 2001 design. It was constructed of Omega 384 low-density blocks manufactured by Burrell Mining Products. When the coating thickness on the faces of the seal and the mortar thickness are included, the total seal thickness was about 41 in.

- Average C-drift entry dimensions at seal location: 18.7 ft wide by 7.3 ft high.
- No hitching was used on this seal.
- This seal was constructed in nearly the same manner as the 40-in Omega block seal constructed by Burrell personnel on March 23, 2006, in crosscut 2 as part of the Test No. 1 Protocol.
- The seal was constructed on the 8-in-thick reinforced concrete floor in C-drift (no foundation).
- Before placement of each Omega block, an approximately 0.25- to 0.5-in-thick layer of properly mixed Quikrete BlocBond (1225-51) was applied to the concrete floor where the seal was to be constructed (Figure B-14).
- Construction of the first row (front course, outby) consisted of eleven 8-in by 16-in by 24-in Omega blocks (16-in block dimension parallel with ribs) and two Omega blocks cut to fit. Construction of the first row (back course, inby side) consisted of eight 8-in by 16-in by 24-in Omega blocks (24-in block dimension laid rib to rib) and one Omega block cut to fit. Each block joint consisted of a full wet-bed construction of at least 0.25-in thickness of BlocBond (Figure B-15). Full Omega blocks were laid against each rib; the cut blocks were fitted inside the row.
- BlocBond was used to fill any remaining vertical gaps between the Omega blocks as completely as possible and also the gaps between the Omega blocks and ribs. As each Omega block course was completed, BlocBond was also coated on the exposed inby and outby sides of the blocks and forced into the outside vertical joints and gaps as completely as possible.
- The remaining 10 Omega full-block courses were installed in a manner similar to the first course, except BlocBond was applied to the block joints and all surfaces of the block before placement and the blocks in each subsequent course were alternated to stagger the block joints. The last full course of blocks on the outby side of the seal was not positioned until the inby side was completed to the mine roof (which included wedging the boards).
- On the 12th and final course, each block was cut and mortared into place on the inby side of the seal to leave an approximately 2.5-in gap between these blocks and the mine roof. A row of 1-in-thick by 8-in-wide by 10-ft-long hardwood rough-cut boards with a thin top coating of BlocBond was then set in BlocBond across the flush end of the inby side of the seal. Wedges on 6- to 12-in centers were then driven perpendicular between the boards and roof and flush with the inby side of the seal. BlocBond was used to fill any gaps between the wedges and to cover all exposed wood. Working from the outby side, BlocBond was thrown by gloved hand along the top of the seal (between the top

block course and the mine roof) to fill the gap between the board on the inby side and the center of the seal. A row of rough-cut hardwood boards (1 in thick by 8 in wide by 10 ft long) was placed (rib to rib) on top of the center section of the seal and wedged tightly into place. BlocBond was placed on top of the middle row of boards before installation across the top of the seal. BlocBond was again thrown with gloved hands along the top of the seal to fill any gaps between the wedges of this middle row of boards and the mine roof.

- The last course of full block (11th course) and the cut block of the 12th course on the outby side were then mortared into place (Figure B-16). Additional BlocBond was then thrown by gloved hand to fill the 2.5-in gap between the top block course and the mine roof from the middle row of boards to where the board flush-mounted on the outby side was positioned. Once this gap was filled as completely as possible with BlocBond, a final row of 1-in-thick by 8-in-wide by 10-ft-long rough-cut boards was placed across the top of the seal and flush with the outby side of the seal. (No attempts were made to pack the BlocBond across the top of the seal using any tools or implements other than throwing the material by gloved hand, as shown in Figure B-17.) BlocBond was placed on top of these boards before installation across the top of the seal. Wedges were driven perpendicular to the board on 6- to 12-in centers with the wedge ends flush with the outby side of the seal. BlocBond was used to fill any gaps (between the wedges) and to cover all exposed wood. Using gloved hands and trowels, an approximately 0.25-in-thick coating of BlocBond was then applied to the outby and inby sides of the seal.

- Approximately 237 Omega blocks from the Bluefield, WV, plant were used to construct this seal.

- Ninety-seven bags of Quikrete BlocBond high-strength fiber reinforced surface bonding cement (1225-51; 50-lb average bag weight) were required to complete this seal, including that used as joint mortar and face sealant (approximately two bags of BlocBond were required to coat each face of the seal). Not including water weight, the total weight of BlocBond was approximately 4,850 lb. In general, each batch of mortar and sealant used for this seal consisted of two bags of BlocBond mixed with approximately 4.25–4.50 gal of water. This calculates to approximately 206–218 gal of water used with the 97 bags of BlocBond required for constructing this seal.

- Construction of the seal took approximately 9 hr (36 worker-hr). This does not include the time required to spot the construction materials to the site.

Figure B-14.—A 0.25- to 0.5-in-thick layer of properly mixed BlocBond on the floor.

Figure B-15.—Omega blocks laid wet with fully mortared (BlocBond) horizontal and vertical joints.

Figure B-16.—Completing the 11th and final full block course on the outby side.

Figure B-17.—Hand slinging the BlocBond to fill the gaps between the previously installed center board and the outby board.

4. Sago 40-in Omega Block Seal Design in Crosscut 3
Construction Date: July 5, 2006
NIOSH–MSHA–WVOMHS&T Seal Testing - Test No. 3 Protocol

On July 5, 2006, a Sago nominal 40-in-thick Omega block seal was constructed by personnel from Ki Corp. (a NIOSH contractor) in crosscut 3 between B- and C-drifts in the LLEM. It was constructed of Omega 384 low-density blocks manufactured by Burrell Mining Products. When the coating thickness on the faces of the seal and the mortar thickness are included, the total seal thickness was about 41 in.

- Average crosscut dimensions at seal location: 18.8 ft wide by 6.75 ft high.
- No hitching was used on this seal.
- The seal was constructed approximately 6–7 ft into the crosscut (as measured from the C-drift side) on a small concrete foundation that tapered from 0 to 3 in thick on top of an 8-in-thick reinforced concrete floor designed to assist in the leveling of the first course of blocks.
- The entire crosscut (roof, ribs, and floor) was dampened with a fine water spray just before starting the construction. An approximately 1.5-in-thick dry layer of Quikrete BlocBond (1225-51) was applied by hand on the floor (concrete foundation) where the seal was to be constructed (Figure B-18). This dry layer of BlocBond (10 bags) was then dampened with a fine water spray to a wet, but not runny condition (Figure B-19). The water spray was applied as uniformly as possible across the entire layer of dry BlocBond for 4 min (~11.6 gal of water).
- The Omega 384 blocks were manufactured at the Bluefield, WV, plant.
- Construction of the first row (front course, C-drift side) consisted of thirteen 8-in by 16-in by 24-in Omega blocks (16-in block dimension parallel with C-drift) and one Omega block cut to approximately 8 in to fit. Construction of the first row (back course, B-drift side) consisted of nine 8-in by 16-in by 24-in Omega blocks (24-in block dimension parallel with B-drift) and one Omega block cut to approximately 3.5 in to fit. A 0.25-in gap was desired between each block; however, these gaps varied from 0.25 in up to as much as 0.5 in between the blocks in this course and subsequent courses because of the nonuniform size of the blocks. The undersides of the Omega blocks were dampened with a light spray of water before positioning the blocks onto the dampened layer of BlocBond that was previously laid on the floor. No BlocBond was applied to any of the block joints before positioning the blocks onto the floor. The entire first course was laid in the same manner.
- Using gloved hands, BlocBond mixed with water (according to the manufacturer's recommendations) was then applied to all accessible surfaces of this first course of Omega blocks. BlocBond was used to fill the (vertical) gaps between the Omega blocks as completely as possible by gloved hand and also the gaps between the Omega blocks and ribs. As each Omega block course was completed, BlocBond was also coated on the exposed B- and C-drift sides of the blocks and forced into the outside vertical joints and gaps as completely as possible using only gloved hands.
- The remaining eight full courses of Omega blocks were installed in a manner similar to the first course except the Omega blocks were not dampened on the underside. All of the Omega blocks were first laid in each course, then the BlocBond was applied in the same

manner as the first full course (Figure B-20). The remaining rows were alternated to stagger the joints.

- On the 10th and final course, each block was cut as necessary to result in a 2.5-in gap between this course and the mine roof. For the blocks on the B-drift side, the Omega blocks were cut to approximately 1.5 in by 4 in by 16 in and pushed through from the C-drift side. Mortar was then applied by gloved hand from both the B- and C-drift sides. From the C-drift side, the Omega blocks were cut to approximately 2.75 in by 5.25 in by 24 in. Mortar was then applied by gloved hand from C-drift. The gap between the 10th course and the mine roof was approximately 2.5 in.

- Approximately 200 full Omega blocks and 16 cut blocks were used to construct this seal.

- Three rows of 1-in-thick by 6-in-wide hardwood rough-cut boards were run lengthwise (between ribs) on the top course of Omega blocks. One row of wood was placed across the middle section of the seal, and one row of wood was placed on the flush end on each side of the seal.

- One 1-in-thick by 6-in-wide by 16-ft-long and one 1-in-thick by 6-in-wide by 3-ft-long rough-cut board were placed rib to rib on top of the center of the seal, working from the C-drift side. Wedges were driven parallel to the board between the board and the roof. BlocBond was then thrown along the top of seal and spread by gloved hand to fill any gaps (between wedges) and to cover all exposed wood. At times, one installer used a wedge to push the BlocBond to the centered board. Then one 1-in-thick by 6-in-wide by 16-ft-long and one 1-in-thick by 6-in-wide by 3-ft-long rough-cut board were placed rib to rib on top of the of the seal on the inby side (C-drift). Wedges were driven parallel to the board and between the board and the roof (Figure B-21). BlocBond was then thrown along the top of seal and spread by gloved hand to fill any gaps (between wedges) and to cover all exposed wood. Once this gap was filled as completely as possible with BlocBond, a final 1-in-thick by 6-in-wide by 16-ft-long and one 1-in-thick by 6-in-wide by 3-ft-long rough-cut board were placed rib to rib across the top of the seal and flush with the outby side (B-drift) of the seal. (Other than an occasional use of a wedge by one installer to assist him in getting the BlocBond to the roof gap and pushing some material away from the edge, no other attempts were made to pack the BlocBond across the top of the seal using any tools or implements other than throwing the material by gloved hand.) Wedges were driven parallel to the board between the board and the roof. BlocBond was again thrown along the top of the seal to fill any gaps between the wedges and to cover all exposed wood. Using gloved hands, an approximately 0.25-in-thick coating of BlocBond was applied to the C-drift and then the B-drift sides of the seal.

- Ten bags of dry BlocBond (50-lb average bag weight) were required to provide the 1.5-in-thick rib-to-rib layer (~44 in wide) before laying the first block course. Sixty-seven additional bags of BlocBond were required to complete this seal, including that used as joint mortar and face sealant (approximately two bags of BlocBond were required to coat each face of the seal). Not including water weight, the total weight of BlocBond was approximately 3,850 lb. In general, each batch of mortar and sealant used for this seal consisted of two bags of BlocBond mixed with approximately 4.25–4.50 gal of water. This calculates to approximately 142–150 gal of water used with the 67 bags of BlocBond required for constructing this seal.

- Construction of the seal took approximately 8 hr (32 worker-hr). This does not include the time required to spot the construction materials to the site.

Figure B-18.—Applying the 1.5-in-thick layer of dry BlocBond to the dampened concrete floor in X-3.

Figure B-19.—Applying a fine water spray to the dry BlocBond layer.

Figure B-20.—Applying the properly mixed BlocBond to the first course of Omega blocks.

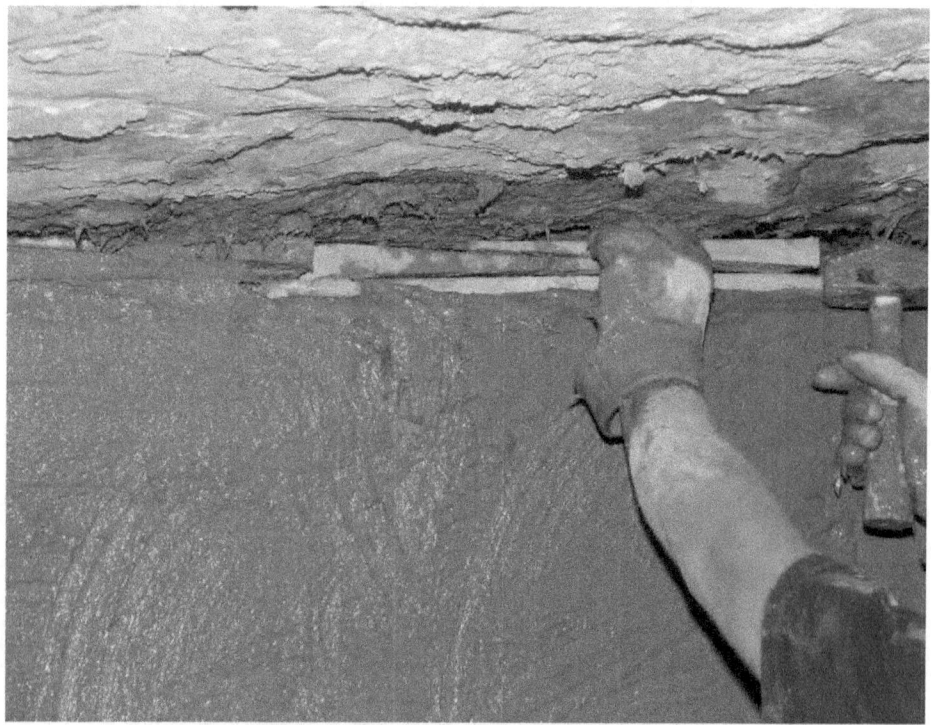

Figure B-21.—Driving the wedges (skin to skin) between the crossboard and mine roof.

5. Sago 40-in Omega Block Seal Design at 320 ft Outby C-Drift Face
Construction Date: July 7, 2006
NIOSH–MSHA–WVOMHS&T Seal Testing – Test No. 3 Protocol

On July 7, 2006, a Sago nominal 40-in-thick Omega block seal was constructed by personnel from Ki Corp. (a NIOSH contractor) in C-drift of the LLEM. It was constructed of Omega 384 low-density blocks manufactured by Burrell Mining Products. When the coating thickness on the faces of the seal and the mortar thickness are included, the total seal thickness was about 41 in.

- Average entry dimensions at seal location: 18.7 ft wide by 7.3 ft high.
- No hitching was used on this seal.
- This seal was constructed in nearly the same manner as the Sago 40-in Omega block seal that was installed in crosscut 3 on July 5, 2006.
- The seal was constructed on the 8-in-thick reinforced concrete floor in C-drift (no foundation) at approximately C-320.
- The entire section of entry (roof, ribs, and floor) was dampened with a fine water spray just before starting the construction. An approximately 1.5-in-thick dry layer of Quikrete BlocBond (1225-51) was applied by hand on the floor where the seal was to be constructed; this dry powder foundation was approximately 48 in wide (Figure B-22). This dry layer of BlocBond (12 bags) was then dampened with a fine water spray to a wet, but not runny condition (Figure B-23). The water spray was applied as uniformly as possible across the entire layer of dry BlocBond for 5 min (~14.5 gal of water).
- The Omega 384 blocks were manufactured at the Bluefield, WV, plant.
- Construction of the first row (front course, outby side) consisted of eight 8-in by 16-in by 24-in Omega blocks (16-in block dimension parallel with outby side) and one Omega block cut to approximately 21 in to fit. Construction of the first row (back course, inby side) consisted of thirteen 8-in by 16-in by 24-in Omega blocks (24-in block dimension parallel with inby side) and one Omega block cut to approximately 4 in to fit. A 0.25-in gap was desired between each block; however, these gaps varied from 0.25 in up to as much as 0.5 in between the blocks in this course and subsequent courses because of the nonuniform size of the blocks. The undersides of the Omega blocks were dampened with a light spray of water before positioning the blocks onto the dampened layer of BlocBond that was previously laid on the floor. No BlocBond was applied to any of the block joints before positioning the blocks onto the floor. The entire first course was laid in the same manner (Figure B-24).
- Using gloved hands, BlocBond mixed with water (according to the manufacturer's recommendations) was then applied to all accessible surfaces of this first course of Omega blocks (Figure B-25). BlocBond was used to fill the (vertical) gaps between the Omega blocks as completely as possible by gloved hand and also the gaps between the Omega blocks and ribs. As each Omega block course was completed, BlocBond was also coated on the exposed outby and inby sides of the block and forced into the outside vertical joints and gaps as completely as possible using only gloved hands.
- The remaining nine full courses of Omega blocks were installed in a manner similar to the first course except the Omega blocks were not dampened on the underside. All of the Omega blocks were first laid in each course (Figure B-26), then the BlocBond was

applied in the same manner as the first full course. The remaining rows were alternated to stagger the joints.
- On the 11th and final course, each block was cut as necessary to result in a 2.5-in gap between this course and the mine roof. For the blocks on the outby and inby sides, the Omega blocks were cut to approximately 5.5 in by 24 in by 16 in. Mortar was then applied by gloved hand from both the inby and outby sides.
- Approximately 216 full Omega blocks and 41 cut blocks were used to construct this seal. *NOTE:* At times, multiple pieces were used from the same cut blocks.
- Three rows of 1-in-thick by 6-in-wide hardwood rough-cut boards were run lengthwise (between ribs) on the top course of Omega blocks. One row of wood was placed across the middle section of the seal, and one row of wood was placed on the flush end on each side of the seal.
- One 1-in-thick by 6-in-wide by 16-ft-long and one 1-in-thick by 6-in-wide by 2-ft-long rough-cut board were placed rib to rib on top of the center of the seal, working from the outby side. Wedges were driven parallel to the board between the board and the roof. BlocBond was then thrown along the top of seal and spread by gloved hand to fill any gaps (between wedges) and to cover all exposed wood. At times, one installer used a wedge to push the BlocBond to the centered board. Then one 1-in-thick by 6-in-wide by 16-ft-long and one 1-in-thick by 6-in-wide by 4-ft-long rough-cut board were placed rib to rib on top of the of the seal on the inby side. Wedges were driven parallel to the board between the board and the roof. BlocBond was then thrown along the top of seal and spread by gloved hand to fill any gaps between the wedges and to cover all exposed wood. Once this gap was filled as completely as possible with BlocBond, a final 1-in-thick by 6-in-wide by 14-ft-long and one 1-in-thick by 6-in-wide by 4-ft-long rough-cut board were placed rib to rib across the top of the seal and flush with the outby side of the seal (Figure B-27). (Other than an occasional use of a wedge by one installer to assist him in getting the BlocBond to the roof gap and pushing some material away from the edge, no other attempts were made to pack the BlocBond across the top of the seal using any tools or implements other than throwing the material by gloved hand.) Wedges were driven parallel to the board between the board and the roof. BlocBond was again thrown along the top of the seal to fill any gaps (between the wedges) and to cover all exposed wood. Using gloved hands, an approximately 0.25-in-thick coating of BlocBond was applied to the inby side and then the outby side of the seal.
- Twelve bags of dry BlocBond (50-lb average bag weight) were required to provide the 1.5-in-thick rib-to-rib layer (~48 in wide) before laying the first course of blocks. Seventy-eight additional bags of BlocBond were required to complete this seal, including that used as joint mortar and face sealant (approximately two bags of BlocBond were required to coat each face of the seal). Not including water weight, the total weight of BlocBond was approximately 4,500 lb. In general, each batch of mortar and sealant used for this seal consisted of two bags of BlocBond mixed with approximately 4.25–4.50 gal of water. This calculates to approximately 165.5–175.5 gal of water used with the 78 bags of BlocBond required for constructing this seal.
- Construction of the seal took approximately 9.5 hr (38 worker-hr). This does not include the time required to spot the construction materials to the site.

Figure B-22.—Applying a 1.5-in-thick layer of dry BlocBond on the dampened concrete floor across C-drift.

Figure B-23.—Wetting the dry layer of BlocBond with a fine spray of water.

Figure B-24.—Positioning the Omega blocks in a staggered pattern for the first course.

Figure B-25.—Applying BlocBond by shovel and gloved hands to the first course.

Figure B-26.—Positioning the second course of Omega blocks in a transverse pattern.

Figure B-27.—Installing the crossboard on the outby side of C-drift seal.

6. Mitchell-Barrett Solid-Concrete-Block Seal in Crosscut 3
Construction Date: September 11–15, 2006
NIOSH–MSHA–WVOMHS&T Seal Testing – Test No. 6 Protocol

On September 11–15, 2006, a 16-in-thick Mitchell-Barrett solid-concrete-block seal with an interlocked center pilaster was constructed by personnel from Ki Corp. (a NIOSH contractor) in crosscut 3 between B- and C-drifts in the LLEM.

- Average crosscut dimensions at seal location: 18.6 ft wide by 6.8 ft high.

- The seal was constructed approximately 6–7 ft into the crosscut (as measured from the C-drift side) on a small concrete foundation that tapered from 0 to 3 in thick on top of an 8-in-thick reinforced concrete floor designed to assist in the leveling of the first course of blocks.

- The crosscut (roof, ribs, and floor) at the seal location was washed using a garden hose just before starting the construction.

- The 6-in by 8-in by 16-in solid concrete blocks used for constructing this seal were purchased in April 2002 from Klondike Block & Masonry Supplies, Inc., Uniontown, PA. These blocks were stored in the LLEM.

- The Type S mortar was packaged in 70-lb bags manufactured by Brixment (purchased in August 2006 from Stone & Company Concrete & Builders Supplies, Connellsville, PA). Each batch of mortar consisted of two parts masonry sand and one part Type S mortar. This Type S mortar and sand mixture was then mixed with water according to the manufacturer's recommendations to obtain the proper consistency.

- The Type S mortar mix (~0.375-in-thick bed) was applied to the concrete floor as each block was laid.

- The dry solid concrete blocks were laid in the wet Type S mortar mix to begin seal construction. Using full wet-bed construction, the Type S mortar mix was applied to all vertical and horizontal dry block joints (Figure B-28). The vertical and horizontal joints were nominally 0.375 in. The blocks were laid in a transverse pattern.

- Construction of the first row (front course, C-drift side) consisted of thirteen 6-in by 8-in by 16-in blocks (16-in block dimension parallel with C-drift) and two partial blocks cut to fit within 0.5 in of each rib. Construction of the first row (back course, B-drift side) was similar except the blocks were offset to result in a staggered joint pattern (to the previously laid block). A total of 26 full blocks and 4 partial blocks were required to complete this first bottom course. The blocks were laid in a similar manner for courses 3, 5, 7, 9, 11, and 13 (although some courses required fewer partial blocks to complete the closure to the rib).

- For the second course, 27 full blocks and 2 partial blocks were installed with the length of the block parallel to the crosscut ribs. Courses 4, 6, 8, 10, and 12 were laid in a similar manner (although some courses required only one partial block to complete the closure to the rib).

- The blocks used to construct the 16-in by 32-in pilaster were interlocked to the 16-in-thick main wall at the center of the seal. The 32-in pilaster dimension was oriented in the C-drift to B-drift direction.

- On the 13th and final course, each block was cut and laid to result in a gap of approximately 1–2 in between this top course and the mine roof (Figure B-29).

- The gap between the top course of blocks and the mine roof was completely filled with mortar throughout the entire width and length of the seal (Figure B-30). No wedges were used in the construction of this seal.
- Approximately 364 full blocks (6 in by 8 in by 16 in) and 23 partial blocks were used to construct the seal. Six half-blocks (4 in by 8 in by 16 in) were used at the mine roof.
- Simulated rib and floor hitching was used on this seal. The hitching was simulated by bolting 6-in by 6-in by 0.5-in thick steel angle (ASTM A-36) to both ribs on each side of the seal and on the floor on each side. This angle was secured by 1-in-diam by 9-in and 12-in-long Hilti Kwik bolts III spaced at approximately 18-in centers. A total of 22 bolts were installed on each side—four 12-in-long bolts on each rib and fourteen 9-in-long bolts on the floor. The steel angle on the floor was installed in three sections on each side of the seal. A 16-in section was anchored to the floor against the pilaster (one bolt on each end of this angle section), and two ~103-in sections were anchored against the seal on the floor to either side of the pilaster (six bolts on each section). Type S mortar mix was used to fill any gaps between the steel angles and seal and the steel angles and ribs (Figure B-31).
- 14.5 bags (1,015 lb) of the Type S mortar (subsequently mixed with sand and water) were required for construction of this seal, and an additional two bags (140 lb) of Type S mortar (mixed with sand and water) were used to fill in the gaps between the steel angles and the seal and between the steel angles and the strata interface.
- Both faces of the seal were subsequently coated with an approximately 0.25-in coating of Quikrete B-Bond, four bags of B-Bond on each side.
- Construction of the seal took approximately 22 hr (79 worker-hr). This included approximately 23 worker-hr to install the steel angle on the ribs and floor on both sides of the seal. This does not include the time required to spot the construction materials to the site.

Figure B-28.—Full wet-bed construction on all horizontal and vertical block joints.

Figure B-29.—Installing cut blocks on top course.

Figure B-30.—Completely filling the gap between the top course of blocks and the mine roof with mortar. The blocks were not wedged.

Figure B-31.—Mortar filling the gaps between the steel angle hitching and the blocks along the floor and ribs.

7. Sago 40-in Omega Block Seal Design Using the Blocks From Sago Mine at 320 ft Outby C-Drift Face
Construction Date: September 21, 2006
NIOSH–MSHA–WVOMHS&T Seal Testing – Test No. 6 Protocol

On September 21, 2006, a Sago nominal 40-in-thick Omega block seal using the Omega blocks from the Sago Mine was constructed by personnel from Ki Corp. (a NIOSH contractor) in C-drift of the LLEM. It was constructed of Omega 384 low-density blocks manufactured by Burrell Mining Products. When the coating thickness on the faces of the seal and the mortar thickness are included, the total seal thickness was about 41 in.

- Average entry dimensions at seal location: 18.7 ft wide by 7.3 ft high.
- No hitching was used on this seal.
- The seal was constructed on the 8-in-thick reinforced concrete floor in C-drift (no foundation) at approximately C-320.
- This seal was constructed in the same manner as described in the Test No. 3 protocol for the Sago 40-in-thick Omega block seal that was installed across C-drift (C-320) on July 7, 2006.
- The entire section of entry (roof, ribs, and floor) was dampened with a fine water spray just before starting the construction. An approximately 1.5-in-thick dry layer of Quikrete BlocBond (1225-51) was applied by hand on the floor where the seal was to be constructed. This dry powder foundation was approximately 48 in wide. This dry layer of BlocBond (12 bags) was then dampened with a fine water spray to a wet, but not runny condition. The water spray was applied as uniformly as possible across the entire layer of dry BlocBond for 5 min (~14.5 gal of water).
- The Omega 384 blocks were delivered from the Sago Mine. The blocks were manufactured at the Bluefield, WV, plant.
- Construction of the first row (front course, outby side) consisted of thirteen 8-in by 16-in by 24-in Omega blocks (16-in block dimension parallel with outby side) and one Omega block cut to approximately 21 in to fit. Construction of the first row (back course, inby side) consisted of eight 8-in by 16-in by 24-in Omega blocks (24-in block dimension parallel with inby side) and one Omega block cut to approximately 4 in to fit. A 0.25-in gap was desired between each block; however, these gaps varied from 0.25 in up to as much as 0.5 in between the blocks in this course and subsequent courses because of the nonuniform size of the blocks. The undersides of the Omega blocks were dampened with a light spray of water before positioning the blocks onto the dampened layer of BlocBond that was previously laid on the floor. No BlocBond was applied to the any of the block joints before positioning the blocks onto the floor. The entire first course was laid in the same manner.
- Using gloved hands, BlocBond mixed with water (according to the manufacturer's recommendations) was then applied to all accessible surfaces of this first course of Omega blocks. BlocBond was used to fill the (vertical) gaps between the Omega blocks as completely as possible by gloved hand and also the gaps between the Omega blocks and ribs. As each Omega block course was completed, BlocBond was also coated on the exposed outby and inby sides of the blocks and forced into the outside vertical joints and gaps as completely as possible using only gloved hands.
- The remaining nine full courses of Omega blocks were installed in a manner similar to the first course, except the Omega blocks were not dampened on the underside. All of

the Omega blocks were first laid in each course, then the BlocBond was applied in the same manner as the first full course. The remaining rows were alternated to stagger the joints.

- All of the undamaged Omega blocks delivered from the Sago Mine were used. Part of the installation of the 10th course required the use of newly purchased Omega blocks from the Bluefield, WV, plant on the inby side of the 10th course. These new blocks were also used to install the 11th course.
- On the 11th and final course, each block was cut as necessary to result in a 2.5-in gap between this course and the mine roof. For the blocks on the outby and inby sides, the Omega blocks were cut to approximately 5.5 in by 24 in by 16 in. Mortar was then applied by gloved hand from both the inby and outby sides.
- Approximately 221 full Omega blocks and 33 cut blocks were used to construct this seal. *NOTE:* At times, multiple pieces were used from the same cut blocks.
- Three rows of 1-in-thick by 6-in-wide hardwood rough-cut boards were run lengthwise (between ribs) on the top course of Omega blocks. One row of wood was placed across the middle section of the seal, and one row of wood was placed on the flush end on each side of the seal.
- One 1-in-thick by 6-in-wide by 16-ft-long rough-cut board was placed rib to rib on top of the center of the seal, working from the outby side. Wedges were driven parallel to the board between the board and the roof. BlocBond was then thrown along the top of seal and spread by gloved hand to fill any gaps between the wedges and to cover all exposed wood. At times, one installer used a wedge to push the BlocBond to the centered board. Then one 1-in-thick by 6-in-wide by 16-ft-long rough-cut board was placed rib to rib on top of the seal on the inby side. Wedges were driven parallel to the board between the board and the roof. BlocBond was then thrown along the top of seal and spread by gloved hand to fill any gaps between the wedges and to cover all exposed wood. Once this gap was filled as completely as possible with BlocBond, a final 1-in-thick by 6-in-wide by 16-ft-long rough-cut board was placed rib to rib across the top of the seal and flush with the outby side of the seal. (Other than the occasional use of a wedge by one installer to assist him in getting the BlocBond to the roof gap and pushing some material away from the edge, no other attempts were made to pack the BlocBond across the top of the seal using any tools or implements other than throwing the material by gloved hand.) Wedges were driven parallel to the board between the board and the roof. BlocBond was again thrown along the top of the seal to fill any gaps between the wedges and to cover all exposed wood. Using gloved hands, an approximately 0.25-in-thick coating of BlocBond was applied to the inby side and then the outby side of the seal.
- Twelve bags of dry BlocBond (50-lb average bag weight) were required to provide the 1.5-in-thick rib-to-rib layer (~48 in wide) before laying the first course of blocks. One hundred additional bags of BlocBond were required to complete this seal, including that used as joint mortar and face sealant (approximately two bags of BlocBond were required to coat each face of the seal). Not including water weight, the total weight of BlocBond was approximately 5,600 lb. In general, each batch of mortar and sealant used for this seal consisted of two bags of BlocBond mixed with approximately 4.25–4.50 gal of water. This calculates to approximately 212.5–225 gal of water used with the 100 bags of BlocBond required for constructing this seal.
- Construction of the seal took approximately 9.5 hr (38 worker-hr). This does not include the time required to spot the construction materials to the site.

APPENDIX C.—AIR LEAKAGE DATA FOR SEALS

Table C-1.—Air leakage measurements before Test 1 (LLEM #501)

Location	Air leakage rates, cfm, at pressure differential of—			
	0.8 in H$_2$O	1.5 in H$_2$O	2.6 in H$_2$O	4.85 in H$_2$O
Seal in crosscut 1	Not evaluated			
Seal in crosscut 2	0	0	0	4.4
Seal in crosscut 3	0	0	4.8	7.4

Table C-2.—Air leakage measurements after Test 1 (LLEM #501)

Location	Air leakage rates, cfm, at pressure differential of—			
	0.8 in H$_2$O	1.5 in H$_2$O	2.5 in H$_2$O	4.9 in H$_2$O
Seal in crosscut 1	Not evaluated			
Seal in crosscut 2	0	0	0	0
Seal in crosscut 3	4.4	7.8	12.2	19.1

Table C-3.—Air leakage measurements before Test 2 (LLEM #502)

Location	Air leakage rates, cfm, at pressure differential of—			
	0.9 in H$_2$O	1.5 in H$_2$O	2.5 in H$_2$O	4.6 in H$_2$O
Seal in crosscut 1	Not evaluated			
Seal in crosscut 2	Use Table C-2 results			
Seal in crosscut 3	Use Table C-2 results			
Seal across C-drift (C-320)	0	0	0	4.8

Table C-4.—Air leakage measurements after Test 2 (LLEM #502)

Location	Air leakage rates, cfm, at pressure differential of—			
	0.85 in H$_2$O	1.55 in H$_2$O	2.55 in H$_2$O	4.55 in H$_2$O
Seal in crosscut 1	Not evaluated			
Seal in crosscut 2	0	0	0	4.4
Seal in crosscut 3	Seal destroyed			
Seal across C-drift (C-320)	Seal destroyed			

Table C-5.—Air leakage measurements before Test 3 (LLEM #503)

Location	Air leakage rates, cfm, at pressure differential of—			
	0.8 in H_2O	1.6 in H_2O	2.4 in H_2O	4.4 in H_2O
Seal in crosscut 1	Not evaluated			
Seal in crosscut 2	Use Table C-4 results			
Seal in crosscut 3	0	0	0	4.4
Seal across C-drift (C-320)	4.8	6.1	8.3	11.7

Table C-6.—Air leakage measurements after Test 3 (LLEM #503)

Location	Air leakage rates, cfm, at pressure differential of—			
	0.85 in H_2O	1.5 in H_2O	2.05 in H_2O	3.5 in H_2O
Seal in crosscut 1	Not evaluated			
Seal in crosscut 2	0	0	0	0
Seal in crosscut 3	0	0	<4.4	5.2
Seal across C-drift (C-320)	4.4	6.5	7.8	12.2

Table C-7.—Air leakage measurements after Test 4 (LLEM #504)

Location	Air leakage rates, cfm, at pressure differential of—			
	1 in H_2O	1.5 in H_2O	2.3 in H_2O	4 in H_2O
Seal in crosscut 1	Not evaluated			
Seal in crosscut 2	0	0	0	0
Seal in crosscut 3	7	10	13.5	17.4
Seal across C-drift (C-320)	9.6	12.6	15.2	17.8

Table C-8.—Air leakage measurements after Test 5 (LLEM #505)

Location	Air leakage rates, cfm, at pressure differential of—			
	0.8 in H_2O	1.5 in H_2O	2.3 in H_2O	4.2 in H_2O
Seal in crosscut 1	Not evaluated			
Seal in crosscut 2	0	0	4.8	7.4
Seal in crosscut 3	Seal destroyed			
Seal across C-drift (C-320)	Seal destroyed			

Table C-9.—Air leakage measurements before Test 6 (LLEM #506)

Location	Air leakage rates, cfm, at pressure differential of—			
	0.8 in H_2O	1.5 in H_2O	2.3 in H_2O	4.2 in H_2O
Seal in crosscut 1	104.4	130.5	182.7	243.6
Seal in crosscut 2	0	0	4.8	7.4
Seal in crosscut 3	17.4	24.4	33.1	43.5
Seal across C-drift (C-320)	0	5.2	8.3	11.3

Table C-10.—Air leakage measurements after Test 6 (LLEM #506)

Location	Air leakage rates, cfm, at pressure differential of—			
	0.6 in H_2O	1.3 in H_2O	2.2 in H_2O	3.9 in H_2O
Seal in crosscut 1	87	130.5	174	226.2
Seal in crosscut 2	0	<4.4	5.2	8.7
Seal in crosscut 3	17	20	29.1	45.7
Seal across C-drift (C-320)	Seal destroyed			

*Delivering on the Nation's promise:
safety and health at work for all people
through research and prevention*

To receive NIOSH documents or more information about
occupational safety and health topics, contact NIOSH at

1–800–CDC–INFO (1–800–232–4636)
TTY: 1–888–232–6348
e-mail: cdcinfo@cdc.gov

or visit the NIOSH Web site at **www.cdc.gov/niosh.**

For a monthly update on news at NIOSH, subscribe to
NIOSH *eNews* by visiting **www.cdc.gov/niosh/eNews.**

DHHS (NIOSH) Publication No. 2009-168

SAFER • HEALTHIER • PEOPLE™

www.ingramcontent.com/pod-product-compliance
Lightning Source LLC
Chambersburg PA
CBHW080254180526
45167CB00006B/2525